ENERGIZING
ORGANIZATIONS

ENERGIZING
ORGANIZATIONS

A NEW METHOD FOR MEASURING EMPLOYEE ENGAGEMENT TO BOOST PROFITS AND CORPORATE SUCCESS

Michael Koscec

Bring new life into your organization!

iUniverse, Inc.
New York Lincoln Shanghai

ENERGIZING ORGANIZATIONS
A NEW METHOD FOR MEASURING EMPLOYEE ENGAGEMENT TO BOOST PROFITS AND CORPORATE SUCCESS

iUniverse books may be ordered through booksellers or by contacting:

iUniverse
2021 Pine Lake Road, Suite 100
Lincoln, NE 68512
www.iuniverse.com
1-800-Authors (1-800-288-4677)

Because of the dynamic nature of the Internet, any Web addresses or links contained in this book may have changed since publication and may no longer be valid.

The views expressed in this work are solely those of the author and do not necessarily reflect the views of the publisher, and the publisher hereby disclaims any responsibility for them.

ISBN: 978-0-595-43185-4 (pbk)
ISBN: 978-0-595-87616-7 (cloth)
ISBN: 978-0-595-87526-9 (ebk)

Printed in the United States of America

This book is dedicated to the memory of Dr. Graham Tucker, a man who was light years ahead in his thinking about the well-being of employees and high energy organizations.

Contents

Acknowledgments

I'd like to thank my wife Debby for her patience and wisdom. Writing much of this book was therapeutic for me. I personally experienced the events that I recount in the book. Some were positive and some were negative. Writing about the negative events forced me to face them again and to relive them. This was difficult and emotional. When I was writing the difficult sections, I thought of several negative titles for the book: Terrorists at Work and The Dark Side of Organizations. Debby helped me to see the positive message in the book. I'm eternally grateful for her wisdom.

I'd also like to thank my many colleagues who accompanied me on my journey and on my quest to more fully understand employee engagement, leadership, and organizational excellence: Dr. Graham Tucker, my mentor and the person who introduced me to values in the workplace back in 1981, long before anyone considered values in business. Graham also introduced me to the book *Servant Leadership* by Robert Greenleaf. This book had a profound effect on my understanding of leadership. Dr. Baba Vishwanath encouraged me to develop models as a way of expressing concepts. Dr. Edgardo Pérez and Bill Wilkerson opened my eyes to the importance of employee mental health as a key factor in understanding organizational health and employee engagement. Dr. Hugh Drouin, whose work in building cooperative communities in difficult government departments is a living testament to what a person of vision, conviction, and faith can accomplish. To some of my clients whose leadership in running exemplary organizations has been an inspiration: Pierre-Yves Julien, Dr. Jacques Messier, Alison DeMille, Cheryl Simpson, Dr. Richard Lu, Dr. Bonnie Neuman, and Dr. Barbara Everett. To my consulting colleagues who are always there to support, challenge, and teach: Lynn Bennett, Hammo Hammond, Dr. Dalton Kehoe, Kelly McCullough, and Dr. Howard Eisenberg. Last but not least, I'd especially like to thank Dr. Don Fulgosi for providing his psychiatric expertise to our work, and for his ongoing insights and encouragement. Don's quick mind, positive attitude, dependability, and friendship will never be forgotten. Finally, I'd like to extend my thanks to Penny Hozy for laboring long and hard editing this book, and to Wilson Santos for designing the book's cover. I extend my most heartfelt gratitude to all these, my fellow traveling companions.

Preface

This book is about bringing new energy into an organization by reinvigorating its employees. The book is divided into two parts. Part 1 contains three sections and is the story of some of the corporate experiences that shaped my thinking about management and leadership. Eventually, these experiences motivated me to form Entec Corporation, the company I established in May 1996. From the beginning, the focus of Entec was clear: to provide the right information in a format that senior management could immediately use to effect change. My corporate experience exposed me to many consultants who conducted employee surveys that did not provide actionable results. What was the point of spending money and asking employees to complete surveys that did not clearly point to follow-up implementation? The driving force behind Entec was to be relevant and to make a difference: give senior management information they needed to reinvigorate their employees and increase corporate success.

Part 1 is based on real events in my life, real people, and actual conversations, but Sections I and III are written as fictional accounts. Section II covers the time in my life between these stories. It was a time of great learning and growth for me, and so I call it "A Time of Learning."

Section I, "Bringing Light into Darkness," is a fictionalized account of events that took place at Ontario Hydro. The story is a record of actual events and conversations. Ontario Hydro no longer exists as a corporate entity, and, for this reason, its identity is not hidden.

Section III, "The Boss from Hell," takes place at Northern Energy, a fictional company. In this case, the company name was changed because the company still exists today as an operating entity. This story is also a record of actual events and conversations. Although both stories are, for the most part, portrayed in real time, some time compression was used.

The reason for writing these stories is to illustrate real life situations that were responsible for causing employee disengagement and poor employee health. In both fictionalized stories, the names of the players have been changed. However, all the players are real. Two names in this book were not changed: Dr. Graham Tucker, one of the most enlightened individuals I've ever met, and Art Bertram, one of my employees who died at work in my presence.

There are forces in the workplace that suck the energy out of employees. These forces impact the employees in many different ways. They can be subtle or they can be highly visible. The subtle forms typically take the shape of non-aligned practices and conflicting processes. They exist, not intentionally, but purely out of management ignorance or neglect. They are not seen but are felt. They include unreasonable demands placed on workers, where the workers have little or no control over workload, scheduling, or deadlines. Workers are in situations where their work objectives are hampered by lack of information and resources. They are boxed in with no support or help. Their working conditions are frustrating and de-energizing. These conditions result in high levels of stress and anxiety that can lead to disease and even death.

The visible forces that suck the energy out of employees can include an uncaring and an unsupportive boss. In the most extreme case, it could be an abusive supervisor where the employee has no course for redress. It could be bullying by the boss or by coworkers. In 2005, the *Globe and Mail* reported that bullying in the workplace had risen 30 percent over the previous five years. Confrontation among coworkers, and between coworkers and their bosses, can also lead to emotional distress and severe loss of personal energy. Personal loss of energy leads to organizational loss of energy. Organizational loss of energy leads to lower corporate performance.

According to Salary.com, employers spend $544 billion a year on salaries for which real work is expected but not being done. The American Institute of Stress estimates that job stress now costs the U.S. economy about $300 billion per year. According to the National Center for Health Statistics, there were 685,089 deaths due to heart disease in America in 2003. The Global Business and Economic Roundtable on Addiction and Mental Health (The Roundtable) estimates that the cost of mental disabilities, including burnout and excessive drinking, cost the Canadian economy approximately $33 billion a year. This number can be multiplied by a factor of ten for the United States, costing the American economy in excess of $330 billion a year. The Roundtable estimates that 20 to 25 percent of the workforce suffers from depression and anxiety disorders at any given time. Research has shown that there is a strong link between depression, low mental and physical energy, and heart disease. In 2002, researchers at the University of Toronto and The Hospital for Sick Children reported a 353 percent increase in prescriptions for antidepressants (from 3.2 million to 14.5 million) between 1981 and 2000. The total expenditures on antidepressants jumped from $31.4 million to $543.4 million.

In Part 2 of the book, we will share the results of our research with clients at Entec Corporation that shows linkages between workplace practices and employee well-being. You will learn that employee engagement is inexorably linked with employee well-being. Organizational scores and employee well-being scores moved up and down in tandem. Strongly performing organizations had high scores for both organizational practices and employee well-being. Poorly performing organizations had low scores for both organizational practices and employee well-being.

In addition, you will learn that some organizational practices affecting employee engagement are the same as the practices affecting employee well-being, while others only affect either engagement or well-being. Another significant finding is that organizational practices only link with mental and physical energy. There are no statistical links between organizational practices and depression or mental focus. The most startling research finding was the complete absence of any statistical links between leadership behavior and any of the three well-being symptoms we measured.

Finally, we will share with you the employee engagement model we developed. The model recognizes that there are two distinct components to employee engagement. The first component is the working environment that is created by local people leaders and by senior managers through corporate-wide policies and practices.

The second component is the psychological makeup of the individual employee. Each employee brings something different to the workplace and will react differently to the same set of stressors. These differences need to be acknowledged and considered when measuring employee engagement. Looking only at the organizational factors provides an incomplete picture. Our model was used to create a unique employee engagement survey that measures both of these components. The results generated by this survey provide insightful data to our clients for follow-up action. The data and recommendations are much more effective in renewing an organization than the data generated by the standard employee engagement or satisfaction surveys that are currently available. The research in Part 2 of the book is based on data that was generated by our clients, using the survey developed at Entec. The data was gathered over a period of six years, involving over twenty-five thousand employees.

I'm eternally grateful to the "brave" clients who were willing to use our employee engagement survey. They were willing to take a risk by using a survey that asked employees some personal questions. We have clients who were drawn to Entec by our unique work. However, on several occasions, two weeks prior to survey launch, clients asked us to remove the emotional wellness portion of the survey. They felt that it was inappropriate to ask their employees questions related to their emotional wellness. We complied with our clients' request. This led us to modularize our initial survey into three different surveys to meet the unique needs of our clients.

It was interesting to see the reaction of employees in those organizations where the "full" employee engagement survey was applied. We discovered that employees found the emotional wellness portion interesting. They had no reservations about completing this portion of the survey. In fact, they were eager to learn more about themselves.

It is my sincere hope that you will find this book entertaining and informative. The book was written to increase your knowledge of and sensitivity to the dynamic forces that can make or break an employee. I hope you will use the information in this book to help create a workplace where employees will be fully engaged and where they will want to become partners in achieving corporate success. I hope you will use this information to breathe new life and light into your organization, so that everyone will win: shareholders, customers, and employees.

Introduction

I have been a student of management for many years: first as a manager and then a senior executive in the corporate world for some twenty-five years, and, for the past ten years, as a management consultant. Over this period of time, I devoured books on management by such well-known authors as Peter Drucker, Ron Zemke, Peters & Waterman, Ken Blanchard, William Ouchi, Robert Greenleaf, and by less well-known authors, Scaff and Fassel, and Kets de Vries and Danny Miller.

I remember how much I loved the book *Built to Last* by Gerry Porras and Jim Collins. When I heard that Jim Collins had written another book, *Good to Great*, I immediately rushed out and bought it. I read it and thoroughly enjoyed it, but I found it unsettling. On the one hand, the book was exciting and motivational, and on the other hand, it depressed me.

I kept thinking about the fact that Collins's research could only unearth eleven companies that made the transition from "good to great" out of his starting list of several thousand. Yet there were thousands of other organizations out there. Eleven companies were a drop in the bucket. The more I thought about it, the more distressed I became. My distress emerged from this stark statistic, and it was also rooted in my twenty-five years of personal experience working inside large, medium, and small organizations, all of which were poorly run and, at their best, were mediocre.

The more I thought about it, the more it became apparent to me that, as much as *Good to Great* is an excellent and valuable contribution to management literature, there was a void in business literature that needed to be filled. What was missing was a book that dealt with the reality of the rest of the struggling horde of poor, mediocre, and average organizations that populate our corporate landscape, and that described the issues faced by the average company, and how the average company solved problems and became better in the process. A book was needed that the average CEO, president, director, or manager could pick up and say, "Yeah, I can relate to that. I've experienced some of the same problems. My company is not a great company, and I'm not a Level 5 or even Level 4 leader, but I'd like to move the bar up just a little bit higher. I would like to do a little better."

This book is for managers at all levels who are good and capable leaders, and about whom no books or articles will ever be written. It is for those who toil each and every day, and who sincerely try to do a good job, but who know their company is just another organization struggling to do the best it can. You might not aspire to be a Level 5 leader, but you do want to make a positive contribution.

This book is also for frontline employees. Employees have more power than they realize. Employees are not victims of their working environment. Hopefully, this book will inspire employees to become active participants with their local people leaders to create a better working experience for themselves and for their coworkers.

Unlike most business books, whose authors have been consultants to Fortune 100 companies and are filled with interesting examples from the biggest and the best, *Energizing Organizations* contains examples from companies that don't aspire to be Fortune 100 companies. Many of the examples in this book are from medium and large organizations. Some are family owned and operated; some are privately owned and run by professional managers; some are public agencies; and some are publicly traded on the stock exchange. But they all have one thing in common: they have CEOs who are intelligent, and who are regular people that no one will ever write about. They are trying to do the best they can with what they have.

Energizing Organizations is for the thousands of CEOs, executive directors, and entrepreneurs of this world who add real value with their ingenuity and their hard work. It attempts to show the power that middle managers have, if only they are prepared to and allowed to exercise it. My hope and wish is that this little book will help make your job easier and your organization more successful.

There are some basic and simple things that can be done to help you create some positive energy and excitement, and, in the process, improve the overall performance of your organization. In this book, I'll share some of my personal experiences and anecdotes in the corporate world. I'll share with you how those experiences shaped my thinking, and how this influenced my personal leadership style and the direction of my career.

I'll also share with you how my colleagues and I developed an employee engagement model that was used to create Entec's Employee Engagement Survey. We view employees and the organization from a holistic-systems perspective. Therefore, our Employee Engagement Survey measures organizational practices and employee emotional well-being in one seamless survey that only takes about ten minutes to complete. Having both organizational and employee information in one database has allowed us to conduct groundbreaking research. This research has raised our understanding of new factors that impact productivity and employee engagement. It has added to our understanding of the critical factors of ethical leadership and the impact organizational practices have on employee performance, absenteeism, and retention, and on employee mental health and well-being.

Why is mental health important? "Employee mental health is the ultimate productivity weapon" is a phrase coined by Bill Wilkerson, President of the Global Business and Economic Roundtable on Addiction and Mental Health. We have heard and read about the fact that we are now in the "knowledge economy" or the "information economy." But I think the "economy of mental performance," another phrase coined by Bill Wilkerson, is a much clearer descriptor of our new economy. Today, our mind is the most important contributor and determinant of economic and competitive success. That's why our research findings provide valuable insight into the impact of organizational practices on employee performance and mental health.

Energizing Organizations will show you how to measure employee engagement in a new way. You will be provided with practical processes to follow up with your survey results that will enable you to create amazing workplaces, where your employees will thrive and be fully engaged. Your employees will win. Your company will win. Your shareholders will win. The material is presented in a simple, straightforward manner that allows you to easily introduce energizing changes into your organization.

To help you on your journey, visit www.EmployeeOnlineSurvey.com. This Web site will give you several different ways to access our unique survey. We are particularly excited by the full-automated option that enables you to introduce the survey quickly and easily into your organization, over the Internet. After you purchase the survey, you will automatically receive pre-survey communication templates that will garner a high response rate. You choose the start and end dates for the survey. When the survey process is completed, press a button and you'll instantly receive your results. You'll be on your way to raising employee engagement and employee health, and in turn, re-energizing your organization.

Part 1

Section I

Bringing Light into Darkness

Preface

"Bringing Light into Darkness" is an account of real events that took place at Ontario Hydro. At the outset, I must make it clear that my own personal working experience at Ontario Hydro was rewarding. Ontario Hydro was good to me. I experienced rapid promotions that recognized my achievements. I was sent on executive training programs at some of the best business schools in America: the Columbia Graduate School of Business in New York and the Wharton School of Business at the University of Pennsylvania in Philadelphia. I was treated well, and I was treated fairly.

The purpose in writing this story is not to denigrate Ontario Hydro. I have no personal axe to grind. My purpose is to illustrate how a toxic workplace can be created, not by malicious intent, but largely by company politics and ignorance. But it is also a story of courage and triumph. I inherited a department whose mode of operation was "How will our actions look politically? How will we appear to the politicians? How will we appear to the public?" Ignorance was largely responsible for creating a very toxic work environment. But ignorance cannot absolve or excuse the creators for the human damage and tragedy that was perpetrated.

The department was responsible for purchasing land for new transmission lines needed to carry power from newly constructed nuclear plants. The land-purchase work was high profile because of its sheer magnitude. It involved over two thousand landowners, thousands of acres of land, and millions of dollars. Ontario Hydro was under a microscope. Everyone was watching to ensure that the landowners were treated fairly. Everyone was watching to see if the land was going to be purchased on time to complete the construction of the new transmission lines.

In this highly charged environment, no thought was given to the property agents who had the responsibility for acquiring the land. They were viewed in the same way as a piece of construction equipment used for building the transmission lines. I'd go so far as to say that the equipment was treated better than the agents. The equipment was treated with respect. It was regularly lubricated and maintained. The agents were neither treated with respect nor looked after. Not surprisingly, the equipment worked well. The property agents began to break down.

Bringing Light into Darkness

In the seventies and eighties, Ontario Hydro was a company with twenty-five thousand employees and annual revenues of $5 billion. It was the second largest electric utility in North America after the Tennessee Valley Authority. Ontario Hydro was pursuing an aggressive nuclear expansion program. This involved the construction of eight new nuclear reactors: four north of Toronto on Lake Huron on the Bruce Peninsula, and four east of Toronto on Lake Ontario at Darlington. New transmission lines had to be constructed to distribute the soon-to-be-generated electricity. New transmission lines need land, and in this case, thousands of acres of land were required.

Electricity is distributed on 500-kilo volt (kV) transmission lines from the power generating plants. These transmission lines need a swath of land that is typically three hundred feet wide. In this case, one line would run from the Lake Huron reactor, south to the main electric transmission corridor, running east-west along southern Ontario and toward Toronto. The transmission lines from the Darlington nuclear station (east of Toronto) would run east from Darlington to Belleville. The job that I accepted made me responsible for the acquisition of this land. It was 1979, and the land acquisition department had a budget of $50 million a year and a staff of thirty-five property agents.

I started my new job on a Monday morning in the middle of May. The reality of the new job set in by Tuesday afternoon. The property acquisition department was beset with problems. Each week, the chairman of the board received ten to twelve executive enquiries, usually complaints about one of the property agents, the settlement offer, or the property acquisition process. They came either directly from a landowner or through their lawyer or local politician. All of these enquiries landed on my desk. A response had to be completed within a week and sent to the chairman. In turn, the chairman's office issued an official reply under the chairman's signature.

This was certainly one way to get immersed in the job. On the one hand, it was firefighting of the worst kind. On the other hand, as much as I didn't like working this way, I very quickly learned about many of the issues facing the department. I immediately set one goal for myself: to eliminate all executive enquiries in one year. This meant going from forty to fifty enquiries each month to zero.

In the process, I learned the following:

1. The reason for all the executive enquiries was an unwritten policy that had been established jointly by my predecessor and my boss. The

essence of this policy was to be as generous as possible with the land-owners in order to expedite the land purchases. "Flexibility" was the operative word. At first blush, one might think this was a good policy. But the policy backfired. The transmission lines had to be built and completed before the completion of the new reactors. To ensure that the necessary lands were acquired in time to construct the transmission lines, all of the lands were expropriated. However, within the provision of the Expropriation Act, as a first step, a landowner was able to negoti-ate the best possible price for their land. If our company failed to reach an amicable deal with a landowner, we could go to arbitration and the arbitrator ruled on the terms of the final settlement.

Most of the landowners were farmers, and farmers are shrewd negotia-tors. It didn't take them long to figure out our company's "flexibility" approach, so they stalled. They realized that stalling as long as possible would ensure the highest price. My predecessor was very politically ori-ented, and he felt that arbitration was not an option. It wouldn't look good politically. So, negotiations became protracted. The farmers used every tactic they could think of to complain about their inability to reach an agreement. They would complain by going to their local mem-bers of the provincial parliament (MPP). A group of neighboring farm-ers would band together and hold out as a group. They would complain directly to the chairman about their mistreatment.

2. The fallout of this unwritten policy of "flexibility" had a devastating impact on the property agents, who were responsible for negotiating the land purchases. The agents were professional land appraisers and negoti-ators. They conducted thorough appraisals of every property, and they knew the fair market value of the properties in their respective areas. They were well trained to determine the "injurious affection" of any possible loss of income that might be incurred by a high-cash-crop oper-ation, such as tobacco. In the case of a business, they were trained to determine the cost of lost income due to the incursion of our activities.

    The problem with this unwritten policy of "flexibility" was that their professionalism was constantly being undermined by their head office bosses (my predecessor and his boss), who responded in favor of the landowner every time they received an executive enquiry. Naturally, this knee-jerk reaction by head office had the effect of increasing the number of executive enquiries. An exception became the rule for negotiating land acquisition deals. In other words, land purchases were mostly being negotiated through the back door. The process completely demoralized the property agents. They were frustrated by the lack of support from head office; morale was low; and any desire to do a good job vanished.

The job of conducting land appraisals was a professional endeavor. A process was followed to determine fair market value. When the appraisals were completed, they were presented to the agent's local supervisor for approval. However, there was another significant issue that the property agents had to contend with. The act of negotiating with landowners is a stressful activity. The stress was elevated by the tight timelines that were imposed. The stress was increased even more by the political nature of the policy of "flexibility," where the agents no longer knew the rules or the guidelines of the job. They were always looking over their shoulder. They never knew when the hammer from head office would come down and undermine their professional efforts. But added to all of that were two other stressors.

1.   Since the bulk of the landowners were farmers, the property agents typically conducted their negotiations with the farmer and his wife at the kitchen table of their farmhouse. Before my arrival, many letters were sent to the CEO from a group of farmers who had banded together to hold out for prices well above fair market value. Over the course of two months, there were several unpleasant incidents experienced by the property agents in dealing with this group of farmers.

2.   The agents spent the whole week away from home living in motels. The combination of job-related stress and being away from home proved to be unbearable for some of the agents. During the day, they spent much of their time in the car, driving from one meeting to another, eating fast food, and drinking coffee. At the end of the day, they ate dinner alone, spent some time at a local bar, and than crashed in their motel rooms with a bottle of liquor, where they completed their daily paper work and watched some television. These were the working conditions of my staff upon my arrival.

In the next portion of this story, I'll recount conversations and events that followed after I started my job. All the events described in this book are true. Only the names of the people have been changed. The conversations accurately reflect the content of the discussions that ensued. However, I've taken the liberty of time compression. For example, the content of two or more conversations is sometimes compressed into one to speed up the description of the situation and to get to the main points. In actual fact, many of the conversations spanned several meetings and telephone calls over a much longer period of time.

The following pages will describe the connections I saw between workplace practices, employee health, and productivity; between leadership, corporate politics, and productivity; and between corporate lawyers, company practices, and productivity.

Week One

Day One: Monday

My first day at work, I was overwhelmed and didn't quite know where to start. I had eight executive enquiries on my desk. I had several purchase agreements waiting for my approval. My first priority, however, was to meet with my boss, and to meet the three managers who reported to me. Meeting the rest of my staff would have to wait for a week or two. I called Lisa, my boss's secretary, and arranged a meeting first thing after lunch. I asked Wendy Morris, my executive assistant, to call my three managers and see when they were free. She arranged for them to come to my office on Tuesday at 9:00 AM. Wendy told me it was easy to arrange the meeting. They dropped everything because they were anxious to meet me.

I walked into Martin Pasternak's office at 1:00 PM that afternoon. Martin was the director of property. He had been with Hydro for some thirty years and had worked his way up the ranks to this position. He had been director for the last seven years. We had had an extensive discussion when I was interviewed for the job.

"Come in, come in, Michael. Please sit down. How are you settling in?"

"I must admit, I'm a bit overwhelmed," I blurted out without thinking.

"That's perfectly understandable. It'll take you a few weeks to get a handle on everything."

"Martin," I said, "I wanted to meet with you so that I'm clear on your expectations. I'd like to get your views on what you think are the most important aspects of this job. I know we talked a little bit about the objectives during my interview, but I'd like to get more specific details, if you wouldn't mind."

"Absolutely," Martin said. "I wanted to have this conversation with you as soon as possible as well." Martin leaned back in his high-backed leather chair and continued. "Your predecessor, Gordon Slazenger, did an outstanding job. You see, we're in a fishbowl. All eyes are on us: our chairman, our president, the minister of energy, all the local politicians spanning dozens of municipalities along the hundreds of miles of the 500 kV lines, not to mention all the affected landowners and some special interest groups. Your job is politically very sensitive, and Gordon did a great job of managing the politics.

"You see," he continued, "we have to complete the land acquisition so that the construction division can finish building the transmission towers and connect the wires. All of this has to be completed before the nuclear plants are up and running. We estimate that the plants will be ready to generate energy in about two years—three years max. This means that all the land should be acquired in eighteen to twenty-four months. We're in possession of over half of the land. Construction is well underway on those portions. However, it has been brought to my attention that the negotiations on the remaining lands are becoming much more difficult and are taking a lot longer to close. So, the pressure is really on us. The pressure is really on you and your staff to get results."

"Have you been using a particular acquisition strategy?" I asked, now more overwhelmed than I thought possible.

"Yes, and I was coming to that. Because we're being so closely watched, we've developed a strategy of flexibility. You see, we expropriated all the lands that we needed to secure our ownership rights. The expropriation gives us the right to go on these properties and start construction regardless of the status of the final purchase agreement. However, politically, we would prefer to be in a position where our construction crews work on a property that has been paid for. We don't want to be perceived as bullies. We want to be seen as a caring company. We got enough bad press during our approval process to build the nukes and the transmission lines. We want to be seen as a good and caring corporate citizen going forward—therefore, our policy of flexibility. Michael, I want you and your team to bend over backwards and give the landowners every benefit of the doubt. Just get the deals done so we don't have to pull the rights given to us under the expropriation approval. You have the budget to do the job. There's nothing standing in your way."

"Thanks, Martin. I really appreciate your insights. They're very helpful."

"No problem, Michael. I'm always here if you need me." Martin stood up and we shook hands. I thanked him again and started back to my office. As I was leaving, he said, "And incidentally, you'll probably be getting several executive enquiries from the chairman's office. You need to treat these on a priority basis. I have a commitment to him that he will have a reply for every enquiry on his desk within one week.

When I returned to my office, I thought about our conversation. In some respects, Martin's description of the job seemed fairly straightforward. Be kind, give the landowners what they want, and seal the deal. I had lots of money in the budget, so money was not an issue. Then I looked at the executive enquiries on my desk. I picked up each one and read through it.

Dear Mr. McMaster:

There are ten tobacco farmers in my municipality who have been negotiating with Hydro for the past six months and they don't appear to be making any headway. They approached me through their lawyer, Charlie Hewitt, and asked me to intervene on their behalf. Is there anything you can do on their behalf to expedite these negotiations?

Sincerely,

The Honourable John Bradford,
Mayor of Newcastle

I read the next.

Dear Mr. McMaster:

I was recently approached by Jorgen Utpott, one of my constituents. He informed me that he is one of the landowners affected by the construction of the new hydro line that is running through my municipality. He said that he is completely frustrated by the negotiation process with your company. He feels that the price being offered is completely inadequate and that the price represents less than 50 percent of the true value of his property. I must agree with his assessment. In this regard, can you please look into this situation on behalf of Mr. Utpott, who is very anxious to put this whole business behind him.

With kindest regards,

Derek Fletcher
Councilor
Municipality of Hastings

The other six letters were similar. All of a sudden, it hit me. If everything was as simple as Martin had described, why were all these letters sitting on my desk? Why were all these landowners complaining about mistreatment if we had a policy of flexibility where we were supposed to give them everything they wanted? I got this sinking feeling in the pit of my stomach. Why was this job so politically charged? What was really going on here that I had not been told?

The process for answering the executive enquiries was to send them to the appropriate land acquisition manager, who would prepare the responses and send them back to me. Since I was meeting with the managers the next morning, I figured these letters would be a good place to start. This would be the beginning of my education and my orientation. I felt that some of the answers must lie in the background to these letters. Before I knew it, it was 6:30 and the day was over. It was time to leave, but I decided I'd better be back the next morning by 6:30 AM. I was going to need every minute to wrap my mind around this new challenge.

Day Two: Tuesday

Al Draper, Gary Tout, and Brian Bandy arrived promptly at 9:00. Al was the land acquisition manager on the west side of the province, responsible for the line running from the Bruce Peninsula to London. Gary was in the middle in charge of the line running through Halton Hills. Brian had the east side. His line ran from the new Darlington nuclear station, east along Highway 401 into Belleville. I had not met these men before.

There was a round conference table in my office with four chairs. I invited them in to sit down. They each had a cup of coffee in their hands and it was obvious they had met at the food court in the basement of our building prior to the meeting. We spent the first few minutes in small talk, and then we got down to business. I began by saying:

"Martin and I spent some time together yesterday. He gave me an overall perspective of the land acquisition situation. I came away with three insights. The first is that we have a policy of flexibility. The second is that our jobs are highly visible and, therefore, highly political. And the third is that we have approximately 45 percent of the land acquisition left, and we have only eighteen months to complete all of the purchases."

I stopped for a moment to see if anyone wanted to jump in and say something. All three were very still and very quiet. I watched their body language for clues, but there were none. They just sat there and looked at me. I continued.

"I can live with pressure and tight timelines. However, I feel a little uncomfortable with the policy of flexibility, and I feel especially uncomfortable with the political part. I was hired for this job because of my property development experience. I've been involved in putting together some very complex property development deals. But I'm not a politician. I don't like politics. I don't know how to be political. So I'm feeling quite uncomfortable."

I stopped again. All three were still sitting straight in their chairs. The expressions on their faces had not changed. They were focused and attentive, and I knew they were sizing me up. Again, no one offered to say anything, so I continued.

"You don't know me so I thought I'd share a couple of things with you. First, I like to manage along the lines of what I refer to as structured freedom. In other words, you know your jobs better than anyone. My job is to support you and to provide you with all of the resources you need to do your job. Do any of you participate in a sport?" Al put his hand up and said very softly, "I play soccer."

"Do you play on a team that's part of a league?" I was so pleased that someone had finally spoken that my question came out sounding overly enthusiastic.

"Yes," came back the quiet reply. I decided not to pursue the soccer conversation because we needed to move forward.

"I see my job being the same as the soccer coach's. I'm here to guide you, to coach you, and to do everything in my power to make sure you succeed. As a part of that, under normal circumstances, I'd sit down with each one of you individually to jointly develop your business objectives for the year. However, because we're working under such tight time frames, I'll have to give you clear, quantifiable land acquisition targets and then ask you to tell me what you need to achieve them. We have to talk about the gaps and barriers that are potentially going to prevent you from reaching your targets, and work together to remove those barriers."

I stopped and saw a look of disbelief and skepticism in their eyes. "What are your thoughts about working together in this way?" I asked.

Still silence. I, too, remained silent. Finally, Brian said, "Yes, it sounds good."

"Okay," I said. "I think the best way for us to start this dialogue is to talk about these executive enquiries that we have to answer in the next few days. Brian, I've got a letter here from the mayor of Newcastle about the tobacco farmers. This is in your area, right? I guess you're familiar with the situation. What's the background to these negotiations?"

Brian shifted in his seat and cleared his throat. "Well, how should I begin? I'm not sure how much you know about tobacco farming. Tobacco is a high-value crop. As such, the farmer has to do everything in his power to ensure a good and healthy crop. To minimize the risk of crop failure, tobacco growers rely heavily on sprinkler systems. The sprinklers ensure that the tobacco plants get the right amount of moisture to grow to maturity."

Brian continued. "The issues in this instance are all related to irrigation. From a safety point of view, how far must a sprinkler system be located from the actual 500 kV wires? Can the existing systems be used and, if not, are there replacement systems that can be used? Is irrigation even allowed under a 500 kV line? Will all this land be lost to tobacco growing? How much land will be lost for maintenance easements? These are just some of the questions.

"We were not able to answer many of these questions," Brian went on. "Gord, our previous manager, got involved in the negotiations because of these complexities. Gord, Sandy Riggs, my agent on this project, and I met with Charlie, their lawyer. Charlie wanted to hire a consultant who had expertise in these matters. The cost of the consultant was going to be $50,000, and Hydro would have to pay. The meeting concluded with Gord saying that he would think about Charlie's proposal, and that he would get back to him in a week. After the meeting, Gord told Sandy and me that he wasn't interested in hiring any consultant. Hiring the consultant was going to drag on the negotiations for another two months, and he wanted the deal done now. He instructed us to make a new offer to Charlie by adding $50,000 to the pot. This brought the total package to approximately $530,000. We prepared a revised offer, and one week later, we presented the new offer to Charlie. I guess Charlie didn't like the offer."

I looked at Brian first, and then at the other two managers. "Do you have any thoughts on how to overcome this impasse? Anybody?"

Al spoke first. "Maybe you should hire the consultant and get some straight answers."

I turned to Brian. "What do you think about that?"

"Yeah, I guess that's a good idea." Since I was close to the open door in my office, I leaned back and called out to Wendy. "Wendy, will you please call Charlie Hewitt and arrange a meeting as soon as he's available? The meeting is for Brian, Sandy Riggs, and me. Thanks."

The rest of the day followed the same pattern. I had lunch brought into the office, and we continued working. As we worked through all of the executive enquiries, the discussion began to flow more freely. I think they began to appreciate the problem-solving nature of our meeting. We were working together as a team. However, I knew it was going to be a while before they began to trust me.

Over lunch, I invited Al, Gary, and Brian for a drink and dinner when we finished. They accepted. Working through the executive enquiries gave me some valuable insights into some of the problems. I saw the drink and dinner as an opportunity to continue to dig deeper and build on the momentum we'd started to develop during the day. It was nearing 4:00 when we finished with the executive enquiries. I mentioned to the three managers that I'd like to have a meeting with all of the property agents as soon as possible. They checked their daytimers, and we agreed to meet on the following Wednesday at 10:00 AM. This would give the agents a chance to drive to Toronto from their disparate locations. I asked Wendy to find a suitable meeting room that could hold forty people, to arrange for sandwiches, and to send out a notice to everyone.

Wendy had made reservations for us at Barbarians Steak House. It was about a fifteen-minute walk from the office and we all needed the fresh air.

Dinner

We arrived at Barbarians about 5:30 and settled in at a table in a quiet corner of the restaurant. We agreed to dispense with drinks and order a bottle of wine. After a couple of sips of the wine, we all started to relax and talk about our families. Brian and Al were married and they each had two kids. Gary was divorced and was living with his partner of the past two years. After we finished studying the menu and ordered our food, I thanked them for their hard work, for their cooperation, and for being so helpful today. I decided that now was the time to probe a little deeper and find the major barriers to concluding their negotiations.

"We had a fairly intense day today, but I feel we accomplished a great deal, and I think I was able to learn a few things. First, I learned that throwing money at a problem does not necessarily help resolve that problem. The tobacco situation was a good example and there were others that we discussed. There was always something else brewing below the surface. Would you agree with that?" There was a boisterous laugh and heads nodding in agreement.

"I also learned that you and all of your staff are very knowledgeable and professional in your appraisal work and in your negotiations. Everyone seems to be dedicated and wants to do a good job." Again there were nods of agreement, but a bit more restrained than before. "What is really preventing you and your agents from doing your work to the best of your abilities?" I ordered another bottle of wine.

It was Al who spoke first. "You're absolutely right when you say we have a professional bunch of guys. The problem is that they haven't been allowed to do their work. They take their job very seriously, and they take a great deal of pride in their work. They're also very responsible. When they put together their offer package, they prepare the best offer possible. They always lead with their best foot forward."

Gary jumped in. "But the problem is our policy of flexibility. Farmers are a very smart lot. They see that the neighbors who refuse to accept a great offer and who hold out for a better offer will typically get it." Now Brian piped in. "It's like everyone is playing a game of chicken. The word is out now. Hydro will not avail itself of the provisions of the Expropriation Act that provides for arbitration, where, if we reached an impasse in our negotiations, we could go to court. The courts would weigh the evidence and determine fair compensation. But Hydro doesn't want to look bad. They don't want to look like a bully."

And now it was Al's turn. "It's more than playing chicken to see who will blink first. The property owners who are left have witnessed the negotiations over the past eighteen months and are, in essence, blackmailing Hydro. They're holding out to extort as much money as they can. They know that Hydro will blink first with an even bigger and better offer. So, they hold out for more—and that's why it's getting so political."

Then the floodgates really opened. "If we can't reach an agreement with a landowner, the landowner will call Martin or Gord, or write the chairman. Martin and Gord will come out to the field, rap our knuckles, and tell us we're not doing our job. They then meet the landowner, with the agent in tow, and give them what they want. The agent then has to prefabricate all of the supporting documentation so that the purchase agreement looks legitimate, while Martin and Gord return to head office. They undermine all of the work we did.

"What was the point of doing the appraisal in the first place? What was the point of spending the time to organize a professional offer package and entering into an honest negotiation? Why bother going through all that when Gord will come to the field and overrule everything we did and force us to create some imaginary package that's not based on any facts? Our agents are responsible and accountable for doing a professional and credible job. Yet they are forced to write fiction. They don't know what to do. We as managers don't know what to do. We never know when the other shoe will drop. Our agents lose all credibility, and the landowners continue playing Hydro like a fiddle."

It was an eye-opener for me and gave me a lot to think about. It also gave me some insight into the inner workings of the policy of flexibility. At that point, I was exhausted and couldn't retain any more information. I stopped asking questions about work. The rest of the evening was more relaxed. We enjoyed our meal and talked about many things, parting company at 9:30.

Although we'd had a long and intense day, a positive momentum was starting to build. I was learning about the business issues, and we were beginning to develop a relationship. At dinner, we'd carried this momentum a little bit further.

I learned more valuable insights, and it gave us a chance to learn more about each other. However, we were still at the talking stage, and I knew it was going to take longer to gain their trust. That trust would be earned through my actions. I was still a long way from proving myself to them.

Day Three: Wednesday

I arrived at work on Wednesday at 7:00. After the previous night's discussion, I decided to arrange a meeting with the chief legal counsel at Hydro to talk about the legal aspects of expropriation and the options that were available for settling negotiation disputes between Hydro and the landowners.

Wendy arrived at work at 8:30, her usual time. I asked her to give me the name and number of Hydro's chief legal counsel. "His name is Arthur Ferguson," she said. "Would you like me to set up an appointment?"

"No, I'd prefer to call him myself."

Luckily, Arthur was in his office when I called. I briefly explained that I wanted to gain a better understanding of the expropriation process and asked if he could meet with me. He agreed. We arranged to meet at 9:00 on Friday.

The next two hours I worked through my in-basket. There were several deals that needed my approval plus a new executive enquiry. Around 11:00, I got a call from Gary, telling me that Steve Roper, one of his agents, had had a stroke. He'd been rushed to the hospital an hour ago. I asked Gary to keep me informed of his situation. The afternoon was booked with my first departmental meeting where I was going to meet my fellow managers in the property division.

Day Four: Thursday

I arrived at the office at 6:30 on Thursday morning. I used the first hour to study the file in preparation for the meeting with Charlie Hewitt, the lawyer who was representing the tobacco farmers. We were meeting Charlie at his office at 11:00. Charlie's office was right in the downtown area of the old village of Newcastle. Brian told me to give myself about an hour and a half to get there. We agreed to meet at Charlie's office at 10:45 so we'd have a chance to talk briefly before the meeting. I decided to leave the office at nine.

I still had an hour and a half to get some more work done. At 8:30, Gary called to tell me that Steve Roper was doing fine. He'd arrived at the hospital in good time after suffering the stroke. Apparently, there wasn't going to be any lasting physical damage. The doctor was planning to release him from the hospital on Friday. However, the doctor said he wanted Steve to stay at home for two weeks. During that time, Steve was to see his doctor once a week. After the last visit, the doctor would confirm Steve's return date. I decided to call Steve that afternoon when I returned to the office. I asked Wendy to have flowers sent to his home.

I enjoyed the drive to Newcastle. It was a perfect day, warm with a clear blue sky. The trees on either side of the highway were bursting with young green leaves. I arrived in Newcastle shortly after 10:30. It took only five minutes to find Charlie's office. I spotted Brian parking his car, but there was no sign of Sandy, the property agent on this file.

"What outcome would you like to get from this meeting?" I asked Brian. He looked at me and said, "I'm not too sure. Maybe I'm just too close to this whole thing."

"Then let's just go in with an open mind and see what the man has to say." As I finished the sentence, Sandy drove up. He parked his car and joined us on the front steps of the building.

We walked into Charlie Hewitt's office, and his secretary said that Mr. Hewitt was expecting us and to go straight in.

Brian made the introductions. Charlie had met Brian and Sandy at previous meetings. I was the new boy on the block and didn't know what to expect. Charlie was a slim man, about six foot two, with dark hair. He could not have been more than thirty years old, yet he possessed a strength and confidence well beyond his years. Although the introduction was pleasant enough, once Charlie sat down at his desk, I sensed some hostility.

I got straight to the point. "Charlie, I'm sure you're aware of the fact that the mayor of Newcastle sent a letter to our chairman in which he expressed your frustration in your negotiations with Hydro. As you know, I'm new to this job. I've come here with an open mind and no agenda. My goal is to walk out of your office at the conclusion of this meeting, knowing that we are one step closer to reaching an agreement on behalf of your clients. So, perhaps the best place to start would be with you. Can you tell us what you think are the key issues in this negotiation?"

Charlie sounded annoyed. "I've already been over this with Brian and Sandy. In fact, just before he left, Gord and I had a discussion on this matter. I told him I wanted to hire a tobacco consultant to do a study of the situation. However, I said Hydro would have to pay for it because Hydro needed answers. Gord said he could offer an additional $50,000 to the overall pot. This would have been the cost of the consultant. That works out to $5,000 per grower. How do we know this is an appropriate amount? We don't. Hydro is very clear on what the issues are and what needs to be done to reach a settlement. You guys just have a hearing problem."

This was not a very encouraging beginning. I kept my eyes on Charlie, not wanting to lose eye contact. "Okay, I appreciate your frustration, and it's perfectly understandable. I discussed this negotiation with Brian, and I read the file. But I'd appreciate hearing it from you so that I'm perfectly clear about your concerns. I promise you, this is the last time you'll have to repeat it."

He sighed and began. "There are ten tobacco farmers. Each one started talking with Hydro in good faith when they were initially approached. Their situation was not a typical purchase of farm property. They grow a valuable crop that is highly dependent on regular irrigation. The amount of water is carefully regulated so that the crop gets the precise amount of water during the various stages of growth. Currently, the easement for the high-tension line cuts right through the middle of each of these farmers' lands." He took a sip of water from a glass on his desk and continued.

"By cutting the tobacco fields in half, the easement has a huge impact on the tobacco-growing operation. What we don't know is if all of the land in the easement will be taken out of production. If that's the case, what is the impact of this on lost revenue due to lost production? What is the increase in the cost of farming now that tobacco fields are no longer contiguous? Or is only part of the tobacco field affected—the part immediately under the wires? How close can the tobacco grower irrigate to the wires? Are there other irrigation technologies that can be used under the wires or close to the wires that would be acceptable to Hydro?

"Each one of these alternatives has real implications on the cost of production and/or on the lost income going forward. Is Hydro qualified to provide answers to these questions and to determine fair and equitable compensation for the impact of the 500 kV line? From what I've seen over the past half year, I think the answer is a resounding no. My clients began to recognize this in the early stages. They started to talk among themselves and decided, since they were facing the same problems, they would band together and negotiate jointly on key principles. Once the key principles were agreed to, they would settle individually because the amount of land that is affected in each case is different. But the issues are the same."

Charlie stopped and shot a glance at Brian and then Sandy, and then rested his eyes on me. There was silence for about thirty seconds. I finally spoke.

"Thank you, Charlie. I appreciate your clear explanation of the problem. I can understand the magnitude and complexity of determining fair compensation. Now, tell me, how would you like to proceed?"

"As I said before, I'd like to hire an expert in tobacco growing and tobacco irrigation. I'd like Hydro to pay for the consultant. I'd also like Hydro to accept the consultant's recommendations. Hydro's acceptance means that your compensation will be based on the consultant's report. I already have the name of a highly respected person in tobacco growing, and we have spoken. He understands the issues and is prepared to conduct the necessary research on the impact of the 500 kV line. Everything hinges on the feasibility of irrigation. Once this is determined, then the other calculations can be made."

I thought for a bit. "Charlie, I think what you're saying makes a lot of sense. Brian, Sandy, what do you think?" Brian and Sandy nodded their heads in agreement. I stood up and leaned across his desk, extending my hand. "Let's move forward. You hire this expert. We'll abide by his recommendations and prepare the offers accordingly. Can you please give Brian his name? When Brian gets back to his office, he'll fax you a short note confirming our verbal agreement. How long will the consultant need to complete his recommendations?"

"He said he'd need about two months. If he can get started in the next week or two, we should have his report by the end of July. I'll call him today and find out. Who would you like me to keep in touch with?"

I looked at Brian. "Sandy is the lead agent on this project. Brian, are you okay if he takes over the reins again, because he's the one who'll have to put together all of the final documentation and present the individual offers?"

Brian responded in the affirmative.

"Then I think we're done," I said. "Charlie, it was a pleasure to meet you, and I look forward to the consultant's report and an amicable conclusion to these negotiations."

Charlie stood up and we all shook hands one more time. For the first time since we'd entered the room, I felt Charlie's hostility disappear.

Driving back to Toronto, I thought that if all the lawyers were as reasonable as Charlie, we might be able to settle most of the tough chestnuts within the year. I later discovered that this was wishful thinking.

Day Five: Friday

I arrived at the office of Arthur Ferguson, chief legal counsel for Hydro, five minutes before our appointment. I reviewed a few short notes that I'd jotted down during the discussion on expropriation with Al, Gary, and Brian on Tuesday. Arthur's secretary led me into his office.

"Hi, Arthur. I really appreciate your seeing me," I said. Once again, I got straight to the point. "As you can imagine, there's a great deal of work to be done in a very short period of time on the land acquisition side of the 500 kV lines. I'm on a steep learning curve about Hydro's policies, practices, and all the personalities that are involved. To be perfectly honest, my head is spinning."

Arthur laughed. "Welcome aboard," he said. "I can imagine there's a lot to learn. It's a big job. How can I help?"

"I was told that we pursue a policy of flexibility in our land acquisition. I don't have any difficulty with that. However, it seems to me that we're now facing a number of stubborn landowners and lawyers, and I'd like to have a better appreciation of the options we have at our disposal."

"I'll try to make this explanation as simple as possible," said Arthur. "When we initially expropriated the land, all landowners were served with the notice of expropriation. Hydro then had ninety days to conduct their appraisals and present each owner with an offer of compensation. Once this legal step was completed, the negotiation process began. There were owners who accepted the first offer, and some who disagreed and then accepted the second or third offer. I guess now we're left with owners who have refused the latest offers and are waiting for something better."

"That's my understanding as well," I said. "But what's the next step for Hydro with those owners who've dug their heels in, and who seem to be holding out for the moon?"

"Within the provisions of the Expropriation Act, your next step is to make one final offer and, if that offer is refused, you can proceed to a board of negotiation or to arbitration. If no satisfactory agreement is reached at arbitration, you can bring forth an appeal before the divisional court. It's my understanding," he said, "that the property division views going to arbitration or to court as a Draconian measure and, to date, they have chosen not to take this route."

I thought for a moment and asked, "Would you be willing to proceed to arbitration if asked?"

"If we were asked, yes, we would go to arbitration."

"Thank you, Arthur. I really appreciate the time you've given me." I returned to my office.

I was to learn later that Hydro had never gone to arbitration because they did not have the resources or the expertise to go to arbitration or to appeal before the divisional court.

I went home Friday night, exhausted. I barely had enough energy to spend a little time with my two daughters, Michelle and Samantha, who were eight and five. Friday nights were always treats nights. I typically stopped at a bakery or a European delicatessen and brought them something special. This Friday, I stopped off at a Tim Horton's donut shop that was on the way home and bought a dozen assorted donuts. I was a little embarrassed, walking through the front door with a box from Tim's, but my daughters were very gracious and politely thanked me, for which I was grateful. My wife, who saw how exhausted I was, smiled and said that dinner was almost ready.

After dinner and after the girls went to bed, I poured myself a scotch and sank into the sofa in our living room. Reflecting on the events of the past week, I felt as if I had been at Hydro for a year.

Week Two

Monday and Tuesday

The first two days flew by. Admittedly, I was still pumped about the new job and the challenges it presented. I thrived on the high energy and the speed. Two interesting notices caught my attention. The first was a notice of a meeting called by our new branch vice president, Sam Horowitz. Sam was a career Hydro employee who had recently been promoted. I had never met him, but I'd heard he was very bright and was highly regarded. Sam was an engineer, like 90 percent of all Hydro employees, and he came from the generating side of the company. I didn't know if this was a positive or a negative. Our branch consisted of the property, finance, human resources, and administration divisions. This was significantly different from electricity generation. The meeting was set for the following Tuesday at 10:00. It was to be with the property division and was one meeting of many Sam was planning with all of the divisions under his purview.

The other item that caught my attention was a request for a meeting from Gary. One of his property agents was getting nowhere with a lawyer representing a landowner whose business was affected by our 500 kV expropriation. I found out later that Brian and Gary had been talking after our meeting with Charlie Hewitt. Since we'd arrived at a first positive step with Charlie, Brian suggested to Gary that I might be able to help in his case as well. My week looked fairly full, so I asked Wendy to see if Gary and the lawyer, Mike Singer, were available Friday morning. She later confirmed that I could meet with them on Friday at 11:00.

I called Brian, Gary, and Al on Monday morning to discuss their thoughts on the agenda for the Wednesday meeting with all the property agents. Although this was to be a "get to know you meeting," I felt a little uncomfortable going into the meeting without some sort of agenda. I asked the managers to talk among themselves first, and to share their thoughts with me. Later that afternoon, Al called me on behalf of the others. They had arrived at a structure for the meeting.

Al spoke in his low voice. "We think the meeting should be split into two parts. The morning section could be devoted to discussing difficult negotiations. Each manager and his property agents can select two or three of their most difficult cases ahead of time. The lead property agent can then describe the nature of the negotiation to the whole group. All the agents would then be asked to brainstorm to come up with the best solution. We'll discuss each case until lunch. After our lunch break, we can discuss work-related issues and conclude the meeting at 3:30."

"That sounds like a great format," I responded, enthusiastically. "I like the idea of discussing difficult negotiations and getting everyone involved in suggesting solutions. Will your agents have enough time to prepare this in just one day?"

"No problem. The tough ones are staring them in the face."

"Okay. All that's good, but I'm a little nervous about spending the afternoon on work-related issues. I'm a little concerned it'll turn into a big bitching session, and I'd like to create positive energy. Maybe we could deal with one or two more property cases and spend the last hour on work-related issues."

Al agreed, but added, "The property agents have never had a meeting together like this. There was one big meeting at the time the project was launched. But it was an information dissemination session where Martin and Gord talked about the project, the timelines, and reviewed the legal aspects of expropriation. The property agents are looking forward to getting together for the first time for what appears to be an opportunity to share ideas and to express their thoughts. I don't think talking about work-related issues will be a downer that'll end the meeting on a negative note."

"Okay. Maybe we can discuss one deal after lunch and get into a discussion of work-related issues after that."

I was looking forward to meeting all of the property agents. I was also looking forward to the problem-solving portion of the meeting. But I still felt some anxiety about the second part of the meeting. A more suspicious person might say that I was being set up. I decided it was best to go with the flow at this stage.

Tuesday was a blur. Meetings, work and, to my surprise, no new executive enquiries.

Wednesday

I kept Martin informed of all the meetings I had, both with my staff and also external meetings with people like Charlie Hewitt. Martin was typically in his office by 6:30 in the morning. He usually picked up a toasted bagel and a coffee at the food court in the lower level of our building and brought it up to his office, so between 6:30 and 7:00 was a good time to catch him. The first time I came to see him this early in the morning, I asked him if he minded being disturbed. Martin, a man in his sixties with a mane of silver hair, was genuine in his response when he said he didn't mind me coming in and chatting. Over the months that followed, I was careful not to abuse this privilege and never met with Martin more than once a week.

As in all large corporations, at Hydro there were detailed office standards for all management levels. This included both the physical size of the office as well as the type and amount of office furniture. As a director, in addition to having a conference table that was larger than mine, Martin also had a living-room-size sofa in his office. One time, when I walked in shortly after 6:30, I found him sleeping on his sofa. I quietly tiptoed out and never mentioned the incident.

When I had arranged the meeting with all the property agents, I asked Martin if he wanted to attend. He thought it was a good idea for me to meet with the agents sooner rather than later, but said he didn't see any value in his attending. He half jokingly said he'd probably only get in the way.

At 9:45, I made my way to the conference room. The property division was located in the original Hydro building on University Avenue. It was attached to the new, gleaming, all-glass office tower immediately to the north, where the larger meeting rooms were located. It took me almost ten minutes to find the conference room. Although it wasn't quite 10:00, the room was full. It was an intimidating sight. Without thinking, and with an outstretched hand, I started introducing myself to the closest person. This seemed to be the easiest way to deal with a room full of strangers who were also my employees. No way was I was going to go into that room and sit at the front table for five minutes. Once I started shaking hands and introducing myself, it seemed best to go through the entire room. I felt a sense of relief when I encountered the familiar faces of Brian, Gary, and Al.

When the handshaking finished, I walked to the front of the room and sat behind the table. To my surprise and consternation, I noticed that there were no other chairs behind the table. I stood up and, catching Gary's eye first, I grinned and motioned to him to sit beside me. Then I signaled to Brian and Al. When the three managers brought their chairs to the front and settled in beside me, each one of them had a big, mischievous grin on his face.

At 10:00 sharp, I stood up. "Good morning and thank you for coming here this morning. I'm glad that Gary, Al, and Brian are able to join me here at the front table." There was just a little bit of laughter. "I thought that it was important for me to meet all of you as soon as possible. These three gentlemen set the agenda for today. Did all of you see it?" There was a show of hands.

"I think Brian, Al, and Gary's idea of having a problem-solving meeting is a good one. It'll be an excellent opportunity for all of us to learn from each other and, hopefully, it will lead to the conclusion of some very difficult negotiations. Before we get started, does anyone have a question or would anyone like to add anything to the agenda?" No hands were raised; no voice spoke up.

"Okay, let's jump right in. I understand that a number of you will be presenting examples of your most difficult negotiations." I looked at the three managers. "Do you have a particular order in which you'd like start?" They nodded in the negative. A small detail I'd forgotten to discuss with them prior to the meeting.

"Who would like to kick things off and volunteer?" To my surprise three hands shot up. This is great, I thought. They're not a bashful group.

"Okay, why don't you begin?" I pointed to a well-dressed man sitting in the second row. "I apologize, but I don't remember your name."

"My name is Arnie Vandermeer. I work with Gary on the Halton Hills line."

I motioned to Arnie to come to the table. "Come up to the front, Arnie, and tell us your story."

"One of my toughest files has to do with a lawyer called Derek Wright," Arnie began. "He's a lawyer in Orangeville, and he represents six landowners. None of these landowners has anything in common other than the lawyer. They're not neighbors, and they don't have the same circumstances. One is a mixed-crop farmer, another is a dairy farmer, and so on. They just have this lawyer in common. Anyway, I refer to him as the ambulance chaser. He smells a potential opportunity, and he shows up at the front door of an unsuspecting target. He'll promise the moon and convince the target that they need him. In this case, his pitch is that Hydro cannot be trusted, that Hydro is dishonest, and that Hydro is out to take advantage of them.

"Anyway, all of these owners received offers from me, and our negotiations were going pretty well until Derek showed up on the scene. He's very confrontational, and all the good will that I had established with these landowners has evaporated. Naturally, I have to go through him now. I can't deal with the owners directly anymore. I guess he's promised the owners the moon. In most of these cases, we were getting close to making a deal, and now Derek is demanding two, three, and in some cases, four times more than our last offer. To add to this problem, the other owners in the area with whom we were getting close to a deal, decided to wait and see what happens with Derek's clients. They haven't engaged Derek, but they've decided to play the wait-and-see game.

"So that's my situation," Arnie concluded. "Does anybody have any good ideas?"

Someone from the back yelled out, "I know a guy who could put Derek out of his misery. He'd make it look like an accident." The room filled with laughter.

I looked at Arnie and said, "Can you tell us what you've tried to do so far?"

"When I couldn't go any further, I asked Gary to meet with Derek."

"Gary, do you want to talk about the meeting?"

"Sure," said Gary, clearing his throat. "Arnie and I met with Derek and asked him if we could review each file individually. Derek immediately jumped on us, saying, 'Come back to me with more money, and until you do, we have nothing else to talk about.' I asked Derek to be reasonable and asked if we could talk about specific issues. Again, he came right back at us and said that the issue is money and the specific issue is that there isn't enough of it. He told us he would talk to us when we came back with vastly improved offers, and not until then. There wasn't much we could do but leave.

"After the meeting," Gary continued, "Arnie and I talked and agreed to see if we could find and justify adding another 5 percent to each offer. Arnie went back to each file, turned over every stone, stretched everything to the limit, and came up with increased offers. They ranged from about 5 percent to 7 percent. A week later, we met with Derek again. The new offers were dismissed out of hand. I picked up one offer, held it up to Derek's face and asked him, specifically, what he didn't like about it. He came back and said, 'Not enough money.'

"I moved closer to Derek, until our faces were no more than a foot apart. 'Tell me, specifically, how much you want for this property,' I said. 'Double it and we have a deal,' he said. 'How can you justify doubling this offer?' I asked him. 'Land values are nowhere near what you're asking and there's no injurious affection. We cannot justify your asking price.' He answered by telling us to find a way. Double the offer, double all the offers, and we have a deal. That was the last discussion we had—a week ago."

There was silence in the room. When the silence was finally broken, a discussion ensued about the ethics of being forced to justify an offer, and signing it, when the numbers were all make-believe. I could feel the frustration in the agents' voices.

Another agent spoke up. "We're between a rock and a hard place. We're supposed to be professionals. If we act as professionals and don't get a deal, head office will come down on us. If we don't act professionally to get a deal, we're putting ourselves at risk because it's our signature that's on the deal."

I thought I'd better intervene at this point. "Okay. Let's get back to Arnie's problem. First of all, are there any others in the room with a similar situation?" A few hands went up.

"Describe your situation, and maybe we'll learn something else that could help resolve Arnie's deal."

Steve Shuster, one of Al's agents, began to speak. "Well, my situation is not quite the same, but it's similar in that it involves a very stubborn lawyer. We're getting closer to a deal, but there's noticeable injurious affection in my case, so this gives me a little bit more flexibility to stretch. We boosted the injurious affection significantly, and that seems to be making the lawyer much happier. I've set up a meeting for next week, so we'll see if he accepts our offer."

"Thank you." I tried to pick up on Steve's train of thought. "So, if you're stretched to the max with the property valuation, you have a lot more discretion with injurious affection. Maybe Arnie should see if there's any possibility of being creative with injurious affection."

There was some reluctant nodding, but it was clear these property agents didn't want to completely lose what little integrity they had left.

The rest of the morning continued in a similar vein. It was tough slogging. There was never a shortage of discussion and a desire to come up with solutions. As in any group, some of the property agents were more willing to comment, while others were quiet.

After lunch, we discussed one more situation. Danny Rose kicked off. "I'm dealing with a large dairy farmer. The 500 kV line runs through the middle of one of his pastures. He claims that the effect of the 500 kV lines will dramatically reduce a dairy cow's milk production. He said he read this somewhere. I also heard that the Russians conducted studies on the effects of high-voltage lines on milk production, and found that milk production was not impacted by cows grazing under high-voltage lines."

"So where are you at?"

"We're not sure what to do with this," Gary chimed in.

"Does anyone have any ideas?"

Sandy stood up. "We're dealing with the technical problems of sprinklers under the 500 kV lines. In our situation, the lawyer representing the tobacco farmers found an expert in this area. He was hired to provide recommendations. Maybe you can find an expert in this field as well."

"Excellent suggestion, Sandy," I said. "Let's you, Brian, and I get together after the meeting and see what we can do along those lines."

I began the next part. "I want to commend you for making the negotiation review a really worthwhile exercise today. I hope you got as much out of it as I did. Thank you for your dedication and professionalism. You should all be proud of the work you're doing. I'm proud of what you're doing.

"Now, let's set aside the landowners and the lawyers, and change horses for the rest of the afternoon. I have a commitment to end at 3:30 sharp because many of you have a long drive ahead of you. So we'll spend the next hour and a half talking about work-related issues. There's no structure per se for the rest of the afternoon, so let me kick it off with a question. Looking at your working environment or working conditions, what would you change to make it easier for you to do your job?"

"We spend most of our time in our cars," said one agent. "Our cars are our offices. We drive to our appointments, and we do a lot of paper work in the car after an appointment, like writing notes in our files for follow-up action. This is okay, but once we get into May, and especially June through September, our cars are unbearable. It would be nice if we could have air conditioning in our cars."

It had never occurred to me to ask about cars. As it turned out, all the property agents were supplied with company cars. These were mid-sized economy cars—your basic model with a radio and not much more.

"We also spend a lot of time in motel rooms," said another agent. "Our motel rooms are our other office, where we do most of our paper work after dinner. We're on our own for a large part of the week. We don't have a place where we can talk, share ideas, and discuss problem deals with other property agents like we did today. We get to talk to our field manager over the phone, and that's it."

"I'm not crazy about living in a motel all week and seeing my family only on the weekend," said one of the younger agents. "I have a young son. He's four years old and I feel really cheated. My wife gives me a pretty hard time about being away all week."

Another hand went up. "I know we're not supposed to talk about our negotiations, but I'm really ticked at getting no support from head office. The only time we hear from head office, they sabotage our negotiations, and tell us to do stuff that we know is not right and we can't support."

I was getting an earful. At this point, I interrupted with a question. "I'm not looking for a confession or anything like that, but I'd like to get a better understanding of how your work lifestyle is affecting you personally. For example, someone said earlier that his wife is giving him a hard time for being away all week. I'd like to know if there's anything else going on. What about drinking? Are guys drinking more than they should?"

Joe Romero, one of Al's agents, began to speak. "I think it's fair to say that most of the guys will have a couple of beers at a bar before dinner, and maybe a couple more with dinner. What happens after that in their rooms, I can't say for sure, but I've heard that some guys always keep a bottle in their room."

"Let me ask you this," I said. "If you went home after work each night, do you think you'd be drinking less than when you're on the road?"

This time it was Frank Cousins, one of Brian's agents, who spoke. "I think that's probably true. I know that when I can go home after work, I may have one beer and that's it. I'm with the kids, or talking to my wife, and there's always something that has to be done."

It was 3:30 and time to close the meeting. "I really appreciate the time we spent together today," I said. "I certainly have a better understanding of the issues you're facing in your property negotiations. Equally as important, is that you've opened my eyes to your working environment. I'm brand new to this organization, so I don't know all the limits of my authority. But I give you my commitment that I'll see what I can do to make your job easier. For example, air-conditioned cars. I agree with you that you should have air-conditioned cars. All I can tell you now is that I hear you, I agree with you, and I'll do everything I can to get you air-conditioned cars. I'll also see what can be done to change some of the other issues you raised today. Drive safely."

As we were parting, Al looked me straight in the eyes and said in his quiet but determined voice, "Don't let us down. You made a commitment this afternoon to improve things. Don't let us down."

Thursday

The next two days were a blur, but Thursday morning, I spent my quiet time between 6:30 and 8:00 reflecting on the Wednesday meeting with the property agents. I also reflected on the events of my first two weeks. I thought about the people I'd met, what they'd said, and how they said it. I thought about what I had learned, and I came to the conclusion that I had enough information to set several key goals for myself. I jotted them down:

1. Protect my property agents from head office interference.

2. Restore their sense of professionalism, self-respect, and confidence in their work.

3. Give them a work environment where they have a greater opportunity for interaction with their colleagues.

4. Create a healthier work environment where they can have a more normal life, where their health and their family relationships are not at risk.

5. Reduce and eliminate executive enquiries within the next year.

I looked at the goals and decided they were a good beginning, but I questioned how realistic they were. It occurred to me that I should share these goals with Martin next week to get a quick reality check. His reaction and feedback would help open my eyes to the barriers that needed to be removed. I realized there could be two types of barriers: personal biases and corporate policy. Just thinking about it made me cringe. Changing personal bias can be every bit as difficult as changing corporate policy.

Aside from the potential barriers, a meeting with Martin would provide an opportunity to enlist his support. But I knew it could be a two-edged sword. If he did support my goals, I'd have my boss on my side. If he didn't support my goals, I'd have a long, uphill battle bringing about any positive change. I just needed to decide when to approach him. I'd get that opportunity, but not the way I expected.

Friday

It was the end of May and another perfect day. Mike Singer's office was located right in the heart of downtown Toronto. Gary and I agreed to meet in the lobby of his office building at 10:45. Since the sun was shining brightly and I had not been getting much exercise, I decided that a walk would do me good. I had just enough time to review the file before I left.

Mike Singer was representing a business owner, Harry's Haulage & Gravel. Harry was a small-time operator. He had one dump truck and a small gravel pit at the back of his property. From the photos, I could see a small, corrugated metal building, truck parts, discarded truck tires, and a couple of old, rusting cars. There was a small portion near the back southwest corner of the property that was dug out. According to the property plan, our 500 kV easement was just skirting the back southeast corner. The actual lines were not on Harry's property—only part of the easement—and this easement was not interfering with his gravel pit operation.

We had originally offered Harry a $90,000 settlement. This included $35,000 for the property and $55,000 injurious affection (i.e., disruption of his business). Mike Singer had refused the offer. Dave Dempster was the property agent. Gary and Dave reviewed the deal and upped the offer to $125,000. This was presented to Mike Singer about a month ago, and he had turned it down again.

When I arrived at Mike Singer's office building, Gary and Dave were already in the lobby. When I saw them, I realized how beneficial Wednesday's meeting had been—I knew very well the kind of challenges we faced. I reminded myself that I had to stay objective and not allow myself to become cynical. So I turned my thoughts to the kind of questions I should ask to get a consensus and resolution.

"What's your assessment of this situation?" I asked, after we had greeted each other.

Dave answered, "I think my biggest problem is that I don't know what Singer is thinking. We really need to probe and find out why he doesn't like the offer."

"Did you ask him why he turned down the last offer?"

"Yes. After I presented the offer, he said he'd call me, but he never did. Five calls later, we finally connected about two weeks ago. At that time, he said the offer was inadequate, and he didn't want to discuss it over the phone. And now, here we are."

"I guess he likes to amass those billable hours," I said, realizing I probably wasn't setting a good example by taking that attitude. But they both laughed in agreement.

Mike Singer was a young lawyer. He looked about my age—thirty-three. He was very reserved and soft spoken.

I began by saying, "Mike, I understand you turned down our last offer for Harry's Haulage and Gravel. I'm here to see if we can break through this impasse. Can you specifically address the offer and tell us which parts you feel are not being adequately addressed?"

There was no response. Mike just looked at me. His face was expressionless. An excellent poker face, I thought. Was this going to be a Mexican stand-off?

I finally broke the silence. "Perhaps I can rephrase the question. Tell me your views on the property valuation portion of the offer. Specifically, what are your thoughts on the comparables that were included in the documentation?" I thought that if we broke up the conversation into small bits, we could start talking about specific issues that would ultimately lead to an agreement.

He continued to stare, until finally, he broke his silence. "The property valuation is too low."

Patience is not my strong suit, and he was already testing my limits. I decided there was no point in asking him to discuss the specifics. "What, in your opinion, would be a fair and equitable property valuation that would close this deal?"

"You're the experts. You tell me," he said.

"Are you telling me that you don't have a figure in mind?" I was doing everything in my power to control myself. I wanted to jump across the desk and pound him.

"Like I said, you're the experts. Come back with a better number."

This was maddening. Even stubborn Derek Wright in Orangeville had finally told Arnie he would settle for twice the amount. At least he'd given Arnie a number to work with. This guy wasn't even willing to say what he wanted!

"Okay. Let's talk about injurious affection." I was through being polite. "How much do you want to settle for?"

He just stared back at me in silence, so I filled in the blanks, "I know. We're the experts and you want us to come back to you with a new number."

"Yes." I couldn't believe he actually said something.

Although I'd kept my cool, I was seething. We all stood up. There was no way I was going to shake this bastard's hand. I gave him a little smile and thought to myself, I'm going to get you. I don't know how yet, but I'm going to get you. I turned and walked out of the office.

Gary, Dave, and I went to a small coffee shop around the corner. I'd have preferred to go to a pub—I really needed a drink after that. I had now personally experienced the frustration my property agents felt on a regular basis.

We each got a coffee and a sandwich. Our meeting had lasted less than thirty minutes. "That was good," I said. "I now feel that I'm part of your club. I just experienced what you guys were talking about on Wednesday." Both Gary and Dave gave me a big smile.

"I guess I was really lucky when I met Charlie Hewitt in Newcastle," I told them. "I remember thinking how gratifying it was to walk out of Charlie's office with a feeling of accomplishment. Driving back from Newcastle, I was thinking how much I was going to enjoy this job because it really wasn't that difficult. You just had to be logical." I shook my head. "Tell me the truth. Are the Charlie Hewitts of this world in the minority, and the Mike Singers and Derek Wrights in the majority?"

"Bingo," came the reply from Gary. "You were gently eased into it with Charlie, and today you were shoved hard into the real world with Mike."

"Where do we go from here?" I asked, somewhat helplessly.

"I don't think we have too much choice," said Gary. "We need to revise the offer. We're being forced into playing this crazy game. What do you think, Dave?" Dave took a sip of his coffee.

"Yeah. We don't have much choice. I'll push up the land value a bit and see what I can do with the injurious affection. Hopefully, the new numbers will satisfy Singer."

"How much do you think you can raise the whole offer?" I asked.

"It's hard to say. I'll have to go back to my documentation and just push the envelope. Don't forget, that offer of $125,000 was already pushing the limits."

"Do the best you can, Dave," I said. "But I'm concerned that we're wasting our time with this guy. I'm not sure a better offer is going to put an end to this charade. It could potentially go on forever."

"I agree," said Gary. "But we have to go back one more time."

Gary drove me back to the office. I was looking forward to going home and pouring myself a nice glass of scotch.

Week Three

Monday and Tuesday

The first day of the week always seems to be the most hectic. It's as if little troublemaking gremlins are busy all weekend, conspiring to make every Monday as miserable as possible. At 8:30, Wendy brought me two new executive enquiries. The rest of the day was a rush. I heard that Steve Roper would be starting back to work on a part-time basis starting tomorrow. He was going to join us for the meeting with Sam Horowitz at 10:00. After hearing all the stories last Wednesday, and my personal experience on Friday, I was beginning to wonder if Steve's stroke had been a result of the job.

Tuesday morning, I walked into the conference room for the assembly with Sam Horowitz, our new vice president. This conference room was twice the size of the room we'd been in last week for my meeting with my property agents. I knew all the managers and directors, but there were employees from the other departments in property that I had not met. I got there about ten minutes before the start of the meeting and noticed that people were congregating by departments. Since my group was the largest, they were easy to spot. I moved toward the back of the group and chose a seat behind Al and his property agents.

Art Bertram was seated directly in front of me. Art was the elder statesman of the property agents. His hair was completely silver, and I guessed him to be in his early sixties. Art had been a property agent for more than thirty years and was highly respected by the other property agents, as well as the managers, both for his experience and for his wisdom. He knew Hydro.

Like Al, Art was soft spoken. He didn't say much, but when he spoke, everyone listened. He was Al's right-hand man. Al relied on Art for the really difficult negotiations. Over the years, Art had passed up several promotions. He liked being a property agent. He wasn't interested in being a manager and was not interested in company politics, even though he knew everything that was going on. Art knew how the games were played and who the players were. He had an uncanny ability to figure out who was after whom and what they were trying to gain. He could accurately predict who was going to win and who was going to lose, and what the outcomes were going to mean.

Sam Horowitz arrived a few minutes before 10:00. With him was Martin Pasternak, our director. As soon as Sam walked in, a hush fell over the conference room. Sam was a new entity. Some had heard of him, but I think the majority of the employees in property had not. I doubt that anyone knew what he looked like, except for Martin.

Martin began to speak. "Good morning. It's so good to see you all in one place and at the same time. As you know, Sam Horowitz was recently appointed vice president of our branch. He started in this role at the beginning of the month. I was delighted when Sam called me and asked if he could spend a few minutes with you. Sam would like to speak very briefly and than open it up for questions and answers. By way of introduction, Sam has been with Ontario Hydro for the past twenty-four years. He started out in the engineering division as a young engineer straight out of university. He transferred to the thermal division, where he held several supervisory and managerial positions. He spent a year in audit division and was promoted to director of the division seven years ago. It's an honor and a pleasure to introduce Sam Horowitz."

After the short applause ended, Sam spoke. "I'm very happy to have this opportunity to speak with you this morning. I want to begin by acknowledging the great work that has been done to date on the land acquisition for our new 500 kV lines. This is one of the most important activities going on at Hydro right now. I understand that about 55 percent of the properties have been secured. I also understand that the remaining 45 percent are also the most difficult. I've been informed by Martin that the project is in good hands and that the job will be completed on time."

Sam talked about several other topics. He gave us an update on the nuclear construction at the Bruce and at Darlington. He informed us that the percentage split between hydro, thermal, and nuclear, once the new nuclear plants were completed, would be 25 percent, 20 percent, and 55 percent, respectively. This marked the first time in Ontario Hydro's history that nuclear production of electricity would exceed that generated from water and coal. He also advised us that the latest projection for energy growth was 8 percent over the next ten to fifteen years. This meant that we would have to start considering new generation in about five years.

Sam talked for fifteen minutes. When he concluded his remarks, he asked for questions from the floor. No hands went up. Martin decided to break the silence. "Based on your remarks about future generation, Sam, what will the impact of this be on the property division?"

"That's a very good question," said Sam. "Once this work is done, I don't see the need for the same level of activity. However, there will always be a need for property acquisition. If we begin to consider new generation in about five years, it would still be some years before the land acquisition activity would reach the current level of intensity."

I'd been so busy working, it had never occurred to me to think about the future. Sam's remark was a little unsettling from an agent's perspective. I discovered after the meeting that the number of property agents had doubled three years ago to work on the 500 kV line acquisition. Many of the younger agents had been hired on a permanent basis from the private sector. In those days, contract hiring was not as commonplace as it is today. Even though Hydro did not have a history of de-hiring employees, I could sense some concern from the agents. Like me, they had all been too busy working to think about the future.

Sandy, one of the younger property agents, stood to ask a question. "If you foresee our property acquisition work dropping off substantially in about eighteen to twenty-four months, what will happen to the property agents?"

"Like I said before," replied Sam, "there will always be property acquisition work. The acquisitions may not be for new lines, but for something else. For example, property will be needed for some of the new, small hydro projects we're planning in the northern parts of the province."

No one stood up to ask another question for what seemed like an eternity. But the silence lingered for no more than thirty seconds when Art Bertram stood up. This must have taken a lot of courage. Art was of the old school and very conscious of organizational hierarchy where, in his mind, a frontline employee did not talk freely with a vice president.

"I don't know if you're aware of the severe pressure that the property agents are under. Why doesn't head office do more to support the agents in the field?"

As Art began to sit down, he suddenly collapsed with a loud crashing noise. As he fell, his arm hit the back of his chair, which fell backwards toward me. Art lay motionless on the floor. I moved the chair out of the way as Al bent over Art. He loosened his necktie and opened the two top buttons of Art's shirt. Then he began to give Art mouth-to-mouth resuscitation. I yelled at the person next to me to call for an ambulance. A perfect circle of people formed around Art and Al. I was on my knees near Art's head, trying not to crowd Al. My eyes scanned the length of Art's body. I noticed that the crotch area and the top portion of his pant legs were wet. At first, it didn't register with me, and then I remembered that when a person dies, the muscles relax and the bladder will empty. I put my hand on Al's shoulder. "He's gone, Al. There's nothing more you can do."

A few minutes later, the ambulance arrived. The paramedics could find no vital signs.

Everyone was in shock. As soon as the paramedics carried Art to the ambulance, Sam announced that the meeting was over and suggested that everyone take a half hour to themselves. Al looked pretty shaken. I asked him if he was all right. He just nodded his head.

The room was slowly emptying when it hit me. "Al, do you know where Art lives?"

"He lives in Georgetown. I've got his home phone number and address in my portfolio."

"We need to go to his house and tell his wife in person. Do you know if she works?"

"No, she doesn't."

"Would you like to come with me?" Again, Al just nodded his head. I suggested that he take a couple of days off. He and Art had known each other for a number of years. They had worked well together and liked each other.

Georgetown is small town about forty miles west of Toronto. Since Al lived somewhere in that area, we agreed to go in separate cars. Al suggested I follow him to Art's house. I was glad we went in separate cars. I found the silence comforting. It was the first time in my life that I'd watched a person die. I'd seen dead people before at a funeral home, but this was totally different. Art had died in my presence, right beside me. I was feeling a bit numb, and I guess I was in shock. What was I going to say to Mrs. Bertram? How was I going to say it? Or should I let Al do the talking?

Al and I arrived at Art's house at 11:30. We parked our cars in front and walked up to the house in silence. Al knocked on the door. A stout lady about the same age as Art opened the door. She recognized Al and, as I introduced myself, she looked at me and said, "Art's dead, isn't he."

We spent about an hour with Mrs. Bertram, drinking tea and talking.

My drive to Toronto was completely different than the one to Georgetown. Instead of numbness and shock, I was experiencing rage and anger. Mrs. Bertram had told us she'd wanted Art to stop working. She was worried about him. She didn't like the pressure he was under and told him he was getting too old for it. Art hadn't said much about work, but occasionally, he'd let something slip about a nasty lawyer or unreasonable landowner. He once said that all professional standards had been tossed out the window and it was like the Wild West where lawlessness ruled. I drove straight home. I couldn't bear going back to the office.

Wednesday

Art's death changed my life. I no longer cared if I worked for a large bureaucratic organization. I didn't care about being political and watching every word I spoke. I didn't even care about my career. I was angry as hell. I had only one thing on my mind. This "lawless" situation that Art had described to his wife had to end, and it had to end now.

I walked into Martin's office at 6:30 on Wednesday morning. I had prepared a list of initiatives that I wanted to move forward with. Martin was eating his customary bagel.

"Hi, Martin. Can I talk to you?"

He motioned me in. "That was a tragic event yesterday. How did it go with Art's wife?"

"She said she'd half expected it because of the pressure he'd been under. Actually, Martin, this is exactly what I wanted to talk to you about." Martin motioned me to sit down. "As you know," I said, "I've had the opportunity to meet with a couple of the really intransigent lawyers. It's become obvious to me that they're playing us, and that we're not going to get anywhere with them. You said we needed to show flexibility in our negotiations. But I'm afraid we're at a stage where this policy is doing us more harm than good. There's a fine line between being flexible and being irresponsible. I choose not to be irresponsible. We have a tool at our disposal that we're not using, and the time has come to use it. I'm referring to arbitration. It's available to us under the Expropriation Act."

Martin let me continue. "I'm convinced that the current rule of flexibility is placing our property agents and Ontario Hydro in an impossible situation. The policy has completely disarmed the agents and taken away their ability to negotiate. They're being held hostage. Maybe Art would have died of a heart attack anyway, but I absolutely believe that his death was hastened by the impossible situation he was in. We need to begin using the arbitration process in those cases where the owners or their lawyers refuse to negotiate in good faith."

Martin studied me for a minute. "Michael, our policy has worked successfully so far. We have 55 percent of the deals done."

"But things are different now. Things have changed. Steve Roper had a stroke last week, and Art Bertram died of a heart attack yesterday."

"You're not trying to tell me that Steve Roper's stroke was a result of the work?"

"All I'm saying is that the work environment is not helping. I can't tell you that Steve's stroke was a direct result of the work, and you can't tell me that his stroke had nothing to do with his work." Martin could have fired me on the spot for saying that, but I didn't care. "The point is that if we're going to successfully conclude the land purchases over the next eighteen months or so, we have to begin to do things differently. We can no longer be pushed around by lawyers who know that there are no consequences for their actions."

"Okay, okay, calm down. Why don't you set up a meeting between our chief legal counsel, you, and me, and we can take it from there. Check my availability with Lisa when she comes in. Actually, never mind. I'll ask her to arrange the meeting and let her know about your availability."

"I really appreciate it, Martin. This will make a big difference. I apologize for coming on so strong, but something snapped in me yesterday. I suddenly knew we had to start doing things differently."

This was early June, and it was an exceptionally hot day. In the late afternoon, Al called me to say that Arnie Vandermeer had quit. "Arnie had an appointment with a farmer and his wife today after lunch," said Al. "He chatted with the owners in their kitchen. Negotiations have not been going well in this area. There are no lawyers involved, but a small group of farmers banded together and decided that no farmer would take action on a property settlement until he checked with the others. They were not only close, but they were also mean spirited. For example, in one instance, the property agent was verbally abused, and in another, the agent was chased off the property by a rifle-wielding farmer and a vicious dog."

"At the end of the meeting," Al continued, "as Arnie was walking toward his car, he noticed something unusual about it. Because it was so hot, he'd left all the windows down. He discovered that his car was filled with fresh manure. A farm hand had used a front-end loader to dump the manure through the open windows. Arnie quit on the spot. He walked away and wants nothing more to do with Hydro or his job."

"You're really getting hit hard, Al," I said. "I'm so sorry. I know Arnie is a good property agent. Listen, I met with Martin this morning, and I asked for his support to begin moving forward with arbitration in the difficult cases. A meeting has been arranged next week on Monday morning with the chief legal counsel to talk about arbitration. Can you talk to Arnie and ask him to hold off on any decision just yet? I'll have a word with him at the funeral, as well. I intend to push for arbitration, which will make his job a lot more tolerable."

"I'll give Arnie a call, but I have my doubts," said Al.

Thursday and Friday

The week came to an end quickly. Art Bertram's funeral was set for Tuesday at 11:00. Visitation was going to take place the day before at the J. S. Jones and Son Funeral Home between 2:00 and 4:00, and also in the evening.

Week Four

Monday

Martin and I walked to Arthur Ferguson's office together. Our meeting was scheduled for 11:00. The timing was good because it would give me ample time to get to Georgetown for the visitation.

As soon as we sat down at the conference table in Arthur's office, Martin started to speak.

"We're coming down to the short strokes in our 500 kV land acquisition project. We've reached an impasse with many owners and lawyers, and Michael feels that the only way we're going to break this impasse is to begin arbitration proceedings with the most difficult cases. I must admit that I agree with Michael. We really are up against the wall in many cases, and we need to do something to break the stalemate. What do you think, Arthur?"

"Well, under the law," said Arthur, "you can certainly use arbitration as an option to break a stalemate. There are specific procedures that are laid out that we have to follow. Arbitration will place additional demands on my legal department, however, and at this stage, I'm in no position to say how much workload will be added."

"If we decide to proceed, what do we need to do?" I remained silent and let Martin do the talking.

"You'll have to send all the owners a new package. This package will contain your last and best offer and all the supporting documentation. You'll also have to advise the recipient that if they don't accept the offer within ninety days, you'll be proceeding to arbitration. I have a form letter for this notification that I'll give you. At the end of the ninety days, you have to pass the files to my department. My lawyers, in consultation with the assigned property agent, will initiate the arbitration proceedings."

"Is there a downside that I should know about?" asked Martin.

"There's no downside from a legal or negotiation perspective. The most obvious downside is the public perception. How will this be received by the landowners and their lawyers? Will this create a political storm or will it largely go unnoticed?"

"Thanks, Arthur. I'd like to pursue arbitration, but first, I'd better meet with Sam Horowitz to see what he thinks. I should also touch base with public relations. In the meantime, send me the form letter to include in our package."

On our walk back to the office, Martin said he would call Sam right away.

After lunch, I left for Georgetown.

Tuesday

I came to the office at my usual 6:30. I wasn't in the office for more than ten minutes when Martin came to see me. He advised me that he had spoken to Sam the previous afternoon and that Sam was completely supportive of proceeding with arbitration. He asked me to start. I couldn't believe it. I hadn't expected such a quick response.

Shortly after 8:30, I called each of my three managers and told them the news. To a man, their reactions were identical. "You're kidding me, right? I don't believe you."

"I'm not kidding. We have the green light."

"Fantastic. We can finally do our job."

I told them about the packages they would have to prepare. When the packages were completed, they had to be delivered to my office. I'd review them and then they were all to be mailed out at the same time by registered mail. They agreed that they would need two weeks to do this. Since this was Tuesday, I asked them to have the packages completed by Friday of the week after next. This gave them a couple of extra days.

I left for Georgetown at 9:45. I thought about Art Bertram and felt sad that I had not had the chance to get to know him. During the funeral service, I couldn't get the picture out of my mind of his body lying on the floor. It seemed like such a waste. Art had two sons in their mid-twenties. There were lots of relatives at the funeral. Al and all of his property agents were there as well. The only positive thought that came to mind during the service was that Art's death might not have been completely in vain. At least the property agents and I knew that we were moving forward with arbitration. Art's tragedy had contributed to the speed with which decisions were being made. Hydro was a large and bureaucratic company slow to make decisions. I was convinced that Martin and Sam were as deeply affected by this tragedy as I was. They had also witnessed it, and they had made their decisions quickly.

At the reception, I spent time with Art's family. Afterwards, I looked for Arnie and found him with a small group of property agents. I joined the group and listened to their conversation. When there was an appropriate opening, I touched Arnie's elbow and said, "Arnie, how are you feeling?"

"I'm all right, I guess."

"I know you've been through a lot this past week; first Art, and then the incident with the car. That's too much for anybody to go through. I'm really sorry."

"The manure in the car was the straw that broke the camel's back," said Arnie. "I just cracked. I was fed up and I'd had it."

"This is hard, Arnie, but may I make a suggestion?" I looked straight into his eyes.

"Sure."

"Can I suggest that you take a couple of weeks of vacation and re-think your decision to quit? I'm sure you've heard from Al that we're going to start going to arbitration. I think you'll find that the work is going to be a lot different from now on. You owe it to yourself to really think this through. I was thinking about Art during the funeral, and one of the things that came to mind was that at least some good came out of his death. It motivated our bosses to make decisions more quickly than usual."

"I'm really pissed with those guys in head office," said Arnie. "Present company excluded. They did nothing while we were being crucified in the field. Now that Art's dead, they start making decisions they should have made a year, two years ago."

"I know it's hard, Arnie, but at least they made a decision, and we need to give them credit for that. Anyway, please think about what I suggested. Take two weeks of vacation. Or, if you prefer, take a month's leave of absence. I'll get the approval. Take the time, think, and decide what's best for Arnie Vandermeer."

"I'll think about it," he said.

The rest of the week flew by. One more executive enquiry—but at least there was only one this week.

Week Five

There were still two things that were nagging at me: no air conditioning in the cars, and the agents working out of motels. Strategically, which issue should I raise first? Regional offices are more important than cars. But if I went for the regional offices first, Martin could argue that the agents would not be in their cars as much and, therefore, they wouldn't need air conditioning. Maybe I'd go for the cars first. It was also a smaller issue. Today an air-conditioned car is not a radical concept. But back in 1979, the electric utility culture viewed air-conditioned cars, especially for a utility employee, as a luxury.

I figured I had nothing to lose. On Tuesday morning, I saw Martin again.

"Good morning, Michael. Our early morning meetings are becoming a ritual." Martin motioned me in with bagel in hand.

"So it seems. I'm glad you're an early riser, Martin. There's something I'd like to talk to you about. As you know, my agents spend all of their time in their cars. In fact, their cars are their offices. The Arnie Vandermeer incident really made me think about their situation. You and I are sitting here in the comfort of an air-conditioned office. But our agents work in cars that are not air-conditioned. It's brutal in those cars in the summer. Can I put together a proposal for you where I examine the cost implications of having air-conditioned cars for our agents? I spoke with our fleet people and, apparently, they're preparing the specs for an order of new fleet vehicles as part of their staggered four-year turnover program."

"Sure. Go ahead and prepare a proposal. By the way, I heard Arnie was thinking of quitting. Is that true?"

"As far as I know, he's taking a couple of weeks holidays right now. He was pretty shaken up by Art's death, and the manure incident put him over the edge."

"He's a good agent and I wouldn't want to lose him. Please, keep me informed."

I was tempted to come back with, "If they had air-conditioned cars, I'm sure he'd stay." But I kept my mouth shut. "Yes. I'll let you know as soon as I hear. In fact, I may have another conversation with him."

I met with the fleet manager and asked him if he could order cars with air conditioning. I also asked him how much more it would cost. He advised me that the additional cost of an air-conditioned car would be $22 per month. I found out that fifteen agents' cars were due for replacement this year, eight next year, and twelve the following year.

The fleet manager explained that Hydro did a bulk RFP every three years for their fleet. The RFP had a provision for inflation/cost escalation that allowed Hydro to replace a third of its fleet each year. This way, the fleet was always made up of cars of different ages. Also, from a cash-flow perspective, this process allowed Hydro to avoid one large expense every three years and maintain a similar expense year after year. I asked the fleet manager if the agents whose cars were not up for renewal could get a new car and add their older cars to the fleet. He said no. Only the agents whose cars were up for renewal could trade them.

I had to move quickly if I was going to get my request in with the current order of cars. I had my proposal ready by Wednesday afternoon. I placed the $22 per agent per month in the context of my $50 million budget. I also talked about productivity improvement, safety (manure dumping), and the difference in appearance between an agent who arrives at a meeting from an air-conditioned car, looking fresh and crisp, versus an agent who arrives from a sweltering car, looking sweaty and crumpled. I said that the agents are the representatives of Hydro, and that the company's image would benefit from agents who arrived at negotiation meetings in air-conditioned cars.

Gary called me shortly after 4:00. Robert Styles had had a stroke and been rushed to the hospital. I couldn't believe it. We'd been making such good progress. When I originally met with my three managers in the first week, I said that I wanted to meet with them once each month. This had been temporarily put on hold because of the flurry of activities. Now I realized that, despite all the activities, the four of us needed to take time out to talk, to think, and to assess. After polling the three managers, we agreed that a mid-afternoon meeting with a possible dinner would be least disruptive. All of them thought they could have their offer packages completed by next Wednesday. They would bring them in personally at 2:00, and we would meet afterwards.

Thursday morning at 6:45, Martin came to my office. He had a sly smile on his face, "Where do I sign? I'm sure you have the necessary documentation prepared."

"How did you guess? I have it right here."

My request was submitted on time. Unbeknownst to me, the fleet manager had been thinking about air-conditioned cars anyway, so my request was not a shock. All the new cars he ordered were air-conditioned. They also had a number of interior upgrades that were not in the old cars. Unfortunately, we could only get our own allotment.

When the new cars became available the second week of August, the property agents were ecstatic. It was like Christmas. Even the agents who would not get their cars until the following years were pleased. They had something positive to talk about.

Week Six

On Wednesday, the managers arrived between 1:00 and 1:30. Martin was in the corridor when he spotted Brian coming in. He was impressed that all the deals were in. He wanted to get involved and offered to help. He suggested that he take the deals that were above my approving authority and that he would have to approve anyway. Gary, Al, and Brian knew which deals were over $250,000. There weren't that many. Nevertheless, these were the deals with the largest amount of documentation, and they would take the longest time to read. I really appreciated the help.

"Incidentally," said Martin. "I received the form letter from Arthur Ferguson that will accompany all the offers. I'll ask Lisa to help Wendy complete the forms for each package and organize the final packages."

"Thanks, Martin," we all chimed in together. Martin left us and we started our meeting. "The first thing I'd like to talk about is Robert Styles. He's doing well. It looks like his stroke was a little more severe than Steve's, but he should be okay." They looked relieved at the news.

"Are there any guys that you think may have a drinking problem?" I asked. "I'm concerned about the agents and the way they live. I'm not on a witch-hunt. I don't want to interfere with anyone's personal life. But one death and two strokes since I arrived is a bit much. I think we need to be proactive in preventing any more unnecessary illness."

Brian started to speak. This time, there was no long period of silence as in our first meeting. They were starting to trust me. So much had happened over the past five and a half weeks: arbitration, air-conditioned cars. For the first time, they had support from head office. It was no longer us versus them. We were part of the same team.

"I can think of two of my guys. I periodically get together with different guys for dinner. Mark James and Adrian Lafontaine typically get blotto when we go out."

Both Al and Gary had similar stories. There seemed to be a common thread. All of the agents they mentioned were the older ones.

"I may be wrong," I said, "but I'm convinced that the job structure is contributing to the heavy drinking and to the ill health we've witnessed in the past few weeks. I'm committed to working toward establishing regional offices, which is something else I'd like to discuss with you today. However, I think we should see what we can do in the short term. I received a call from Don Heard, Hydro's staff psychiatrist, two days after Art died. He called to ask how the agents were coping with the death, and to see if he could help in any way. At the time, there didn't seem to be a need. I've spoken with Don again, however. I briefly described the working conditions of the property agents and asked him if he had any programs that dealt with stress, excessive drinking, and so on. He said he did, and that he'd be happy to meet with the property agents. Would you be agreeable to this?"

"Sure" was the unanimous response.

"Okay. Let me ask Don how he'd like to go about this, and then I'll hand it over to you guys to work directly with him and make the arrangements." They nodded their heads in agreement.

"Let's move on to regional offices. The three of you indicated the first time we had dinner, and this has been repeated in subsequent conversations, that many of the work-related problems would be diminished if you and your agents had an office where you could hang your hat. Another part of this would involve relocating the agents to live closer to their new office and to their work so that they could be home nights, rather than living in motels during the week." Most of the property agents were currently living in the suburbs of Toronto.

"I've got a couple of questions," I continued. "Number one, where would you like to have your offices; and number two, how would your agents react to a move?"

Al was first. "I think my office should be in London. The bulk of our future work will be in a seventy-five-mile radius of London. I think my guys would welcome a move somewhere in and around there."

Brian was next. "I was initially thinking of Cobourg. But the best location for our office would be Belleville. I have no doubt that most of my agents would enjoy living there. The cost of housing is so much cheaper than in Toronto. For the same amount of money, they'll be able to buy a house twice as big."

"Gary, what about you? What do you think?"

"It could be argued that we should locate in Orangeville or somewhere south of Orangeville. My preference would be Mississauga. All of our work is on either side and due north of Mississauga. If we located there, we'd be close enough to the majority of our landowners and their lawyers when they wanted to see us in our office. Also, the Mississauga area provides the agents with a lot of housing options. In fact, some live in that area already."

"No question your agents would be least affected by this reorganization," I said. "There's no question that Al's and Brian's agents are the ones that spend more time in motels. Should we pursue this regional office set-up?"

They were unanimous. "Definitely." "I support it." "Let's go for it."

"Okay. Here's what I need from each of you. Prepare your requirements for office space: amount of square feet, type of space, and location within your area. In fact, if you can come up with actual options, specific location, building, costs, and so on, that would be even better. I'll work on the overall plan, justification, and impact of the scheme on the budget. I'll also begin sowing the seeds with Martin. What kind of timing should we be looking at?"

It was Al who spoke up first again. "I think we'll probably need a month."

"And if we're not ready in a month, that's okay," I said. "This is a big change and we don't need to do it at breakneck speed. So, let's target for a month. That'll put us at the end of July. Let me know if you need more time. In any case, shall we target our next meeting for the end of July? If you can have your information ready by then, this topic will be part of that monthly meeting. Oh, I just thought of something. What are your holiday plans?"

None of them were planning on a vacation in the summer, which meant we could continue with the momentum that was being built.

July

The first two weeks of July were spent finalizing the offer packages. Martin had not lost interest. We worked closely together with legal to make sure the packages were perfect, and they were sent out in the middle of July. After the packages were sent, I started to work on my justification for setting up the regional offices.

During the third week of July, I peeked into Martin's office at 6:45. "Can I bother you for a second?"

"No bother at all. I look forward to your visits because I'm never sure what I'll be getting into next."

"There have been a lot of changes in the last little while," I said. "I wanted to bring you up to date on something. After Art died, I had a long conversation with Don Heard, our psychiatrist, about our property agents and their lifestyle. At my request, Don is going to spend a little time with the agents on stress management and on excessive drinking."

"That's good, Michael. Don is a very knowledgeable guy. I knew the agents drank a fair amount, but I never suspected it was a problem."

"I'm not sure that it's a big problem, but I want to be proactive and prevent it from becoming one. That's why I asked for Don's help. But I view Don's intervention as a band-aid. He's not going to be able to solve the root causes of the heavy drinking. From talking with Don, the fact that many of the agents spend all week living out of motels sets them up for an unhealthy lifestyle. This lifestyle encourages a person to gravitate to a bar after work. Their jobs are stressful. Drinking in a bar after work is more appealing to the older agents than yoga, for example. It's not just the drinking. They can strike up a conversation with someone. They're looking for companionship after a day of negotiation and possible confrontation."

"Yes, I know it's not easy."

"The younger agents only see their little kids on the weekends, and I've heard that some of their marriages are strained. Their family relationships are being seriously tested. We've had three serious health-related incidents over the past six weeks. Unfortunately, they occurred on my watch, but they've been in the making for some time. As an organization, I don't think Hydro is blameless. This may sound like a radical comment, but as a manager, I believe that I've got some responsibility for the well-being of my employees. I see the well-being of my staff in the same way as the construction crew managers view the health and safety of their workers. The construction crews are building the 500 kV towers and they're stringing the lines. There are strict safety standards in place to minimize injuries and the loss of life. They keep statistics of all injuries and, God forbid, deaths.

"Our property agents are no different," I went on. "Suppose an agent is seriously injured or killed in a car accident after they leave a bar on their way to their motel room? Do we not share some responsibility for placing a property agent in that position?

"Could Art's life have been prolonged if he hadn't worked in the type of environment that we created? I can't help but think that we bear some responsibility. I'm not saying all the responsibility, but some. The property agents are grown men and they have to take responsibility for their own health and welfare. Nevertheless, we have responsibility for the work design, and the question we have to ask ourselves is, 'Will that work design produce the best results for us and, at the same time, will it promote the well-being of the property agent?'"

Martin looked at me thoughtfully. "I never really thought of it from that perspective," he said. "I mean, from the health and safety perspective. Their jobs are not dangerous, but you have something when you say that the circumstances of their jobs affect their well-being. So where are you going with all of this, Michael? Where are you leading me?

"I think we should look at the feasibility of reorganizing our field operation to achieve two objectives," I said. "The first objective would be to increase the productivity of our agents; and the second sub-objective would be to structure the field organization in a way that will allow the property agents to spend their evenings at home, rather than in motels."

"I'm open to considering a proposal, but I want to see the numbers. Are the alternatives you're thinking about going to cost more, or will they reduce costs? When push comes to shove, the numbers have to work," said Martin. "So go ahead with the proposal, but I want to see the numbers."

"I agree with you, Martin. The numbers have to work. We'll get them to you." I left the office feeling a little anxious. I was so focused on wanting to do the right thing for the property agents that I hadn't given much thought to the numbers. I called each of my managers and asked them to look at the annual cost of travel, lodging, dinners, and any other costs they could think of that would be reduced or eliminated by the satellite offices, and to compare that to the new cost of office space.

Arnie Vandermeer came back to work. He had taken two weeks holidays and four weeks leave of absence. I found out later that Arnie had called Al several times to talk. He wanted to stay plugged in to see if changes were continuing. He had to satisfy himself that the changes were not just an illusion, but that they were real, and they were taking root. Arnie was also in line for a new car.

August

During the second week of August, fifteen agents came to head office to pick up their new cars. In the past, this was usually a ho-hum event. But this time the agents were pretty excited. Not only were the cars equipped with air conditioning, but they were also considerably nicer. They had more options and better interior trim. This was a real morale booster, and it gave me no end of pleasure to see their reactions.

Also during the second week, Sandy called to tell me he had been contacted by Charlie Hewitt. The consultant's report had arrived and he was going to Charlie's office to pick up a copy. Sandy asked me if I wanted a copy. I said no and told him that he was in charge. Our commitment was to follow the recommendations and prepare the offers based on those recommendations. It was his job to work through the report, prepare the new offers, discuss them with Brian, and take them back to Charlie. There was no need for me to get involved.

By the middle of September, all of the tobacco farmers were settled. When the deals were completed, Charlie called me and invited my wife and me to his home for dinner. I accepted and we had a delightful evening at his house overlooking Lake Ontario.

By the end of August, Al, Brian, and Gary had prepared the packages for their offices. Al found an attractive twelve-hundred-square-foot office on the ground floor of a five-story building on the northern edge of London. Gary found an office with two thousand square feet in a mall development near the intersection of Highway 401 and Erin Mills Parkway. The 401 is a major east-west highway that runs from Windsor/Detroit right through Toronto, and ending up in Montreal. The location was perfect. Brian found a fourteen-hundred-square-foot office in a single-story office mall on a major thoroughfare.

The numbers looked much better than anyone had expected. The combined rental for the three offices was less than half the cost of motels and dinners. Furniture was not an issue because Hydro had a stockpile of assorted office furniture. The annual savings were impressive. However, all of these savings would be eaten up in the first year by the cost of moving the property agents.

I pulled all the numbers together and prepared a comprehensive report for Martin. He was surprised. He had no idea that a regional office structure could provide such impressive savings. The human and productivity benefits were the icing on the cake. Martin gave his approval and the offices were secured. The affected agents began searching for their new homes.

By the end of August, the agents were beginning to receive encouraging signs from property owners and some lawyers from the July mass mailing of offers. However, there was still no feedback from any of the hardcore holdouts.

Also in August, both Steve Roper and Robert Styles returned to work on a full-time basis.

September

There was a wave of positive energy in September. The three managers were setting up the new offices. By the end of September, the offices were fully functional. Houses were put up for sale in Toronto. House hunting was underway in London and Belleville.

By the end of September, many difficult deals were either settled or well on their way to being settled. The threat of arbitration had the desired effect.

Martin and I continued with our periodical early morning chats. But now he visited me more often than I visited him.

October

Mid-October would mark the end of negotiations and the beginning of arbitration for many property cases. In September, I called Arthur Ferguson to see what was needed to proceed to arbitration. We'd sent out close to a hundred packages. I was hoping that at least half of these would be settled before the deadline. This still left a large number of potential arbitration cases. To my surprise and dismay, I discovered that our legal department did not have the manpower or the expertise to proceed. All the lawyers in the legal department were corporate lawyers. None of them was experienced at litigation. And the lawyers who were there were fully engaged.

I worked with Arthur on a plan. He suggested that we engage David King, one of Toronto's best litigators. In the meantime, Art would move as quickly as possible to hire a full-time lawyer who was both a litigator and knowledgeable in real estate.

October turned out to be an interesting month. For example, our ambulance chaser Derek from Orangeville settled on the threshold of the arbitration room. There were many similar cases that were literally settled walking into the arbitration room. Others went to arbitration and one went all the way to court—that was Mike Singer. Our final offer to Mike was $212,000. He rejected it. He still would not divulge an amount for which he was willing to settle. So we went to arbitration. The arbitrator recommended a settlement amount of $150,000. Mike Singer chose not to accept this recommendation. He chose to go to court.

In preparation for court, we decided to go back to the beginning. We reviewed the original appraisal, including all the original justification and calculations for injurious affection. All of this documentation could support a maximum offer of $90,000. This was the original offer that had been made and this was our position going into court. The judge heard all the evidence and ordered Ontario Hydro to pay Mike Singer's client, Harry's Haulage & Gravel, the sum of $90,000. I felt vindicated. I had gotten my revenge. Ironically, I felt a little sorry for Harry. I've often wondered what Mike said to him after our offer of $212,000. Had he even presented the offer to Harry?

As a part of this restructuring, I was able to get approval for a head office position. It was set up as a developmental position for property agents, as well as someone who would assist me in working with the legal department on the many arbitration cases. The new position was a problem-solving position, a support position to my managers and to me on complex issues that needed background research, like the impact of the high voltage lines on a cow's milk production, for instance. It wasn't easy to find a consultant who was available to work on these problems.

I had been impressed with Sandy Riggs from the beginning. He had done an excellent job in closing all the deals with Charlie Hewitt. He had always impressed me with his positive attitude and good problem-solving skills. I asked him if he would be interested in working with me at head office. Sandy accepted. He was a great asset, learning a lot and significantly advancing his own personal development.

Each Wednesday I visited one of the satellite offices, where I met with the manager and his property agents. We reviewed the progress of the negotiations, then followed the review with discussion and problem solving on difficult negotiations. At the conclusion of the day, specific action items were listed, some for the agents and some for me where they needed head office support.

On the fourth Wednesday of the month, the managers came to head office to meet with me. This was an opportunity for the managers to talk about common issues and to learn from each other. We talked about management, leadership, and corporate issues. Sandy attended these monthly head office meetings. He participated fully in the discussion. In addition, Sandy was responsible for recording all of the action items that flowed from the meeting, and then following up to ensure that they were completed.

By the end of October, executive enquiries to our chairman began to slow to a trickle. The property agents' morale was reaching new record highs. All the property agents had settled into their new houses and were home in the evenings after work. There were still many deals to settle, but the ground rules were clear. We were being flexible, but we were not being irresponsible. The property agents' faith in their self-worth was restored.

The offices became centers of activity. Instead of conducting all negotiations with landowners at their kitchen tables, many deals were now discussed and settled in the office. Gone were the days of doing paper work in hot cars or motel rooms after supper. I was tracking the absenteeism numbers. When I arrived in May, absenteeism was running at twelve person days per year. By the end of October, we were down to four person days per year. It was the lowest absenteeism rate ever in the property division.

All of these changes did not occur without a price. By the end of October, only five and a half months into my new job, I could not fit into any of my clothes. I had been a hundred and seventy-eight pounds, with a waist size of thirty-four inches when I started my job in May. Now I was pushing two hundred pounds and my waist size had grown by four inches to thirty-eight. All of my life I had been active in sports and physical fitness. Since starting my job, I'd been so consumed by my work that I'd stopped my regular routine of exercising—and it showed.

November to April

We continued with the monthly meetings. Every Wednesday I drove to a different regional office, where I met with the manager and his property agents. We discussed all the difficult deals and came up with follow-up action plans. Some of the solutions involved a different negotiation tactic, reappraisal, input from an expert, or going to arbitration.

The three managers continued to come to head office once a month to share ideas, talk about negotiation issues, and learn from each other. We talked about corporate news and events. Our monthly manager meetings typically lasted the whole day and were divided into four modules. Each month, one of the modules was set aside for a mini training or learning session.

We agreed ahead of time on a list of topics. Sandy would be responsible for doing the necessary research. Sometimes an outside speaker came to talk to us. Another time it might be a trainer from HR who ran through a short module on leadership, or organization skills, or time management. We also invited managers or directors from various Hydro departments to talk to us. For example, a director from the economics department talked to us about their latest energy forecasts; a director from corporate planning spoke to us about Hydro's future plans for generation; a manager from rates spoke to us about projected rate increases. This was all valuable information that the managers were able to take back to their offices and share with their staff.

By the end of the year, executive enquiries were reduced to a trickle and disappeared altogether in the following few months. Morale soared. Productivity increased. There was a steady stream of deals flowing to my desk for final approval. Most of the personal health-related problems disappeared.

One day in April, Martin came to my office at around 6:45, our usual chat time, and told me he had submitted my name for an internal executive training program. In those days, Ontario Hydro ran its own program that was designed to train, groom, and fast-track up-and-coming managers for senior management positions. Martin told me that he hadn't wanted to say anything to me until he knew the decision of the committee that controlled enrolment into this program. He didn't want to build up my hopes because only 15 percent of recommended candidates were accepted. Late in the afternoon of the previous day, he had learned that his recommendation was accepted, and that I would start on the program in September.

The executive training program lasted approximately eighteen months and comprised a mix of lectures, course work, as well as hands-on experience. The hands-on training involved conducting operational audits of various divisions throughout the corporation. Ontario Hydro maintained a permanent operational audit department, which was staffed by a full-time director, but all of the managers and auditors were there on rotation as a part of this program. The department was designed to fulfill two functions: develop future senior managers and ensure effective operations. The department had teeth because it reported directly to the president of Ontario Hydro.

Naturally, I was thrilled with this opportunity. It would lead to all kinds of future possibilities within Hydro. I'd be moving from property to the operational audit at the beginning of September.

Before my departure, the property division organized a farewell dinner. There were the customary good wishes and congratulations. However, one incident is etched in my mind to this day. Al, the manager of the London office, walked up to me and said in his customary quiet but firm voice, "Koscec, you led us to Mecca and now you're abandoning us." At first I thought he was joking. I didn't fully understand what he was saying. Then I began to realize that he was dead serious. In fact, he was more than serious; he was bloody mad. Al was a gifted manager and a thoughtful and intense individual. He appreciated the changes that had occurred more than anyone. His friend, Art Bertram, had died of a heart attack while Al was giving him mouth-to-mouth resuscitation. Al had a huge emotional investment in the new organizational structure, management processes, and leadership style.

I was completely taken off guard by his remark. I didn't know how to respond. I viewed the dinner as a celebration of past successes and future possibilities. Al saw it as a sad time, the end of something good. I could only say, "Al, all the new systems and processes are in place. I'm not leaving you when the job is half done. Yes, there are lots more deals to complete, but you don't have to fear going back to the way things were. Martin is completely supportive, and he's pleased with the way things are progressing. The legal department has hired a full-time lawyer to work with property on arbitration. You have their full support. Our new way of working is now embedded, institutionalized. It has become part of everyday normal life."

Al wasn't convinced, and the joyous mood of the farewell dinner changed for me. It made me look more critically at all the implications of my actions. I had been so engrossed in my work and focused on new challenges that it didn't occur to me to think about my property agents.

Luckily, as time went on, Al's concerns were not founded. Life did go on in much the same way after I left.

Reflections

1. Lack of leadership support combined with poor work design is a lethal combination that can result in employee death.

2. An untrustworthy and unsupportive supervisor will demoralize otherwise motivated people.

3. In a toxic work environment, physical fitness and spiritual fitness cannot relieve stress over the long term. Eventually, the toxic work environment wins out.

4. As a middle manager in a large organization, you have the power to create a healthy and productive work environment if you have a vision, a plan, and the fortitude to see your plan through to a successful conclusion.

5. When creating change, identify two or three practices that you can get started on immediately that will yield the greatest results. All the other things that need changing will begin to fall into place.

6. Before starting a new job, ask your new boss about the worst possible situations that might be waiting for you. You will discover that they're all waiting for you, plus some you didn't expect.

7. Be a systems thinker. Study and follow the nature of connections and relationships between departments and managers. This will identify personal allegiances, partnerships, and biases. You will soon learn whose support you need or don't need to get your job done.

Comments by Dr. Don Fulgosi: A Medical Perspective

In terms of medical and psychiatric implications, this story raises the whole issue of chronic work stress and its pernicious impact on people's health and productivity, and beyond that, of morbidity and mortality. There is a huge body of (medical) literature on this subject. However, the description of the real-life events and working conditions of the property agents clearly demonstrates the connections between work and health. The property agents' lives epitomize both paradigms of chronic work stress: "high demand—low control—low support," and "effort—reward imbalance," with a high degree of unfairness and social isolation. In vulnerable individuals, this is a prescription for death. Unfortunately, Art Bertram was such an individual and he lost his life.

Section II

A Time of Learning

A Time of Learning

After leaving the property acquisition department at Ontario Hydro, the first couple of months at operational audit felt very strange. I missed the high energy of property. I missed the phone ringing, the problem solving, and the interactions with my staff. Audit was quiet. Each audit was conducted by a team of three: two auditors and one manager. Those of us who came to audit from the junior management positions worked as auditors, and those who came to audit from senior manager positions were assigned as managers or team leaders.

Once I got over the absence of the high levels of activity and energy, I found the next year and a half to be exhilarating. Most of the people in audit were bright and interesting. They were very knowledgeable and much better read than I was. They introduced me to the *Harvard Business Review*. Hydro had several subscriptions, and each month I received and read the latest issue. They also introduced me to a number of new business authors. We researched and read these authors and then exchanged books, which we discussed over lunch or coffee. During this time, I read every management book I could get my hands on. I read as broad a spectrum as I could, including Peter Drucker, William Ouchi, Peters and Waterman, Kenneth Blanchard, Ron Zemke, and many more. The moment a new book on management came out, I read it.

There were other books that were not in the mainstream. *The Addictive Organization* by Scaff and Fassel, and *The Neurotic Organization* by Kets de Vries and Danny Miller, in particular, provided valuable insights into what caused organizations to be the way they were. In many respects, these two books became more relevant to my own research than the others.

A couple of my colleagues in audit, including my team leader Bruce, had started working out at the gym and asked if I wanted to join them. I could stand to lose some weight—like the twenty pounds I'd gained in property—so I said yes. This couldn't have happened at a better time. A couple of months before leaving property, I had had a long overdue physical and my doctor had informed me that both my blood pressure and my cholesterol were high. I respect my doctor because he's not a pill pusher. He gave me a choice: either start exercising and eating less or he would put me on meds to reduce my blood pressure and cholesterol. Since it was the beginning of summer, I told him I'd start exercising and eating healthier. But I wasn't doing such a great job at it, so I jumped at the opportunity to work out on a daily basis.

Four of us would meet in the fully equipped gym in the basement of the head office building at 7:00 each morning and work out for thirty-five to forty-five minutes. The workouts were not particularly strenuous. None of us wanted to push too hard at the beginning because we were all out of shape. We lifted weights, used the Nordic Track cross-country apparatus, and ran on the tread-mill. In early spring, Bruce motivated us to start running outside. This meant meeting at the gym at 6:30 and going outside to run at 7:00. We started by running for twenty minutes and slowly increased our distance over time. By the end of the summer, we were running eight miles every morning.

Ontario Hydro had a corporate training center located in the Hockley Valley, an area about fifty miles northwest of Toronto with rolling hills and lush green valleys. During my time in audit, I spent a great deal of time there attending various management courses that were part of our program. I'd go for a run first thing in the morning, and it became a great time to think about the perfect organization.

Graham Tucker

I was only in audit three months when a flyer on the bulletin board at our church caught my eye. It was an advertisement for a workshop on "Values in Business" being offered by Dr. Graham Tucker. The workshop was the following Saturday at the Aurora Conference Centre. Something stirred inside me and I knew I had to go. On Monday morning, I called the contact number and signed up.

The Aurora Conference Centre is a forty-five-minute drive north of Toronto. There were about twenty people at the workshop. Graham Tucker was a thin man, about six feet tall. He spoke softly and was rather droll, and you couldn't help but listen to every word he said. After we went around the circle introducing ourselves in the usual workshop fashion, Graham introduced himself.

"I began my career in British Columbia," he said, "working for Alcan Aluminum as a mechanical engineer. I was interested in designing and building machines and manufacturing plants. Two of my colleagues were faithful Christians. They introduced me to the full meaning of Christianity and my life was completely changed. I dedicated my life to Christ and, shortly thereafter, I returned to university, obtained my Masters degree in theology, and became an ordained Anglican priest.

"This life-changing experience taught me a profound lesson," he went on. "It taught me the importance of the ministry of the laity. I did not become a faithful Christian due to the institutional church. I became a person of faith due to my association with two fellow engineers. I believed that a faithful laity would be the major strength of the church in the future.

"There are over a hundred thousand people working in the heart of down-town Toronto," Graham told us. "These people have their minds focused on commerce, finance, productivity, marketing, administration, and so on. But the fact remains, they are people with a heart, a soul, a spirit—they are human beings. They have human needs for love, compassion, friendship, respect, community, and a sense of belonging. Who do these people turn to on a Tuesday afternoon when they feel they're going to fall apart from severe anxiety brought on by the pressures of work? Who do they turn to if their boss is being overbearing and unreasonable? If they are members of a church or a synagogue, chances are their minister or rabbi is many miles away, somewhere in the suburbs.

"This is why I started the King Bay Chaplaincy. I wanted to provide a safe haven for workers where they work. We provide teaching, facilitate small groups, and provide support programs that help employees to cope and be fully human. There are dozens of luncheon groups meeting all over downtown Toronto. Some are Bible study groups, others are discussion groups on a variety of faith or spirituality-in-the-workplace subjects. We started Operation Bootstraps, a return-to-work program for unemployed men and women. Plus, there's Discovery, a weekend retreat designed to awaken a person's Christian beliefs and to help them live a Christian life 24/7, not just for one hour on Sunday. We welcome everyone at the KBC, all faiths and all cultures, but we don't hide our Christian roots."

Graham continued. "Meeting all the people that came through the doors of the KBC, and hearing their work-related stories, drove our program design and helped to shape the purpose of the King Bay Chaplaincy:

- To gather Christian business people for mutual support, encouragement, and study in a network of small groups

- To equip people for lay ministry in the workplace, through educational programs

- To serve the needs of individuals seeking healing and wholeness in personal and corporate life, through counseling and seminars

- To establish a positive connection between the church and the workplace, and

- To influence businesses to apply Christian principles and values which further the Kingdom of God in the workplace, through conferences and seminars.

"It is the last point," he said, "to influence businesses to apply Christian principles and values in the workplace, that the KBC is currently hard at work on.

Two years ago, I completed my Doctorate at the University of Toronto in values and business. This workshop on "Values in Business" is founded on my thesis. It's a new program offered by the KBC. Today we're going to explore the connection between values and business."

The workshop began with a discussion and exercises that were based on Graham's first book, *It's Your Life—Create a Christian Lifestyle*. We reflected on the dangers of living a life based on two sets of values: one set in the workplace from Monday through Friday, and another set on the weekend when we're with family and friends and attending worship services. This was followed by a discussion of books such as *In Search of Excellence* by T. J. Peters and R. H. Waterman, and *Corporate Culture* by T. Deal and A. Kennedy. Both of these books discuss the importance of building a strong sense of community within a company as one of the most important ingredients for success.

After a session on the hierarchy of values, we moved to the values of leaders and the impact on an organization of the values of its president. For example, Peters and Waterman, in *In Search of Excellence*, make it clear that the value system of the president is the most important factor in building a spirit that results in success. They write that Thomas Watson, Jr., founder of IBM, said, "I firmly believe that any organization, in order to survive and achieve success, must have a sound set of beliefs on which it premises all its policies and actions. Next, I believe that the most important single factor in corporate success is faithful adherence to those beliefs."

We were introduced to many other books during the workshop. The one that stood out for me was Robert Greenleaf's *Servant Leadership*. In 1982, there was little interest in Greenleaf's book. Today, *Servant Leadership* is often quoted. In the book, Greenleaf writes, "Work exists for the person as much as the person exists for the work. Put another way, the business exists as much to provide meaningful work to the person as it exists to provide a product or service to a customer."

During the workshop, my mind kept racing back to my time in property at Hydro. I realized that I hadn't been aware of what I was doing back then. My focus was on problem solving—get the deals done and improve the working conditions of the property agents. I wasn't thinking about my own values. I wasn't thinking about job design, or about leadership styles, or different forms of management. I had never read books by Peter Drucker, William Ouchi, Kenneth Blanchard, Ron Zemke, and the many other authors who wrote about management and leadership. I was flying by the seat of my pants. But what I discovered at this workshop was that I was flying by the seat of my value pants.

I have a strong Christian faith. I take time each morning for prayer and meditation. The way I approached problem solving in my work was based on my
deep-rooted Christian values. And I learned that these values yield results! This
was a huge awakening for me.

After the workshop, I chatted with Graham to see how I could get involved
with the KBC. The whole area of values in the workplace fascinated me. Getting
involved with Graham would be a perfect fit with my work in audit. Since the
KBC was still a new and growing entity, Graham welcomed my interest. He
asked me if I we could meet during the lunch hour the following week to see
where I could be of most use. We made a date for the following Friday.

At this point, I was well into my first audit. My team was conducting an operational audit of Pickering Nuclear Station #1. This was the oldest nuclear station
in Hydro's portfolio. It comprised four 25-megawatt reactors. Pickering Nuclear
Station #2, with four more 25-megawatt reactors, was nearing completion. It was
located immediately to the east of Station #1. When I returned to the office after
the workshop, my mind was spinning. Was there any way I could apply what I'd
learned to this audit? The audit was highly structured and focused on operations.
I reviewed our audit guidelines and terms of reference but could see no possible
way to insert something new into our work, like values or leadership style. I
briefly chatted with my team leader, Bruce, and he confirmed that it wasn't possible. He could appreciate what I was saying, but this was his first audit as well, and
there was no way he was changing anything. I could understand his position.

The Value Analysis Profile

I met with Graham Tucker on Friday. He told me he'd been thinking about
developing a tool that would measure values in an organization, and he had some
ideas. He asked me if I'd be willing to help him develop this tool, which he had
named the "Value Analysis Profile." I agreed. It sounded really interesting. After
that initial meeting, Graham and I met two times a week for the next couple of
months to develop the profile.

By February, we had completed ten separate modules for the Value Analysis
Profile. We decided to set up Management Support Systems, a subsidiary consulting arm of the KBC. Its purpose was to offer to businesses the Value Analysis
Profile and other diagnostics that Graham had developed. As a first step, Graham
and I met with the president of the Canadian Manufacturers Association, who
thought the profiles were unique. He invited us to speak at the quarterly meeting
of members in April. This was an incredible opportunity for us. We would be

speaking to a room of approximately fifty presidents or vice presidents of the largest manufacturing companies in Canada.

Unfortunately, our presentation was not well received. Suffice it to say, there was stunned silence when we finished. They thought we were from Mars. "What do values have to do with business? Are you crazy?" We left the meeting licking our wounds.

Our foray into the world of consulting was a flop. We approached McKinsey and Company, a forward-thinking management consulting firm, whose consultants had written *In Search of Excellence*. They weren't interested either. We shared the profiles with them, explained how to apply them, and discussed the benefits of using them. They were polite, but not interested in measuring values. Values did not fit into their consulting practice. Graham and I were completely deflated and gave up.

Then another opportunity presented itself. The Hydro audit group was asked if we wanted to add anything to our next management workshop. I jumped at the opportunity and requested a half-hour in which everyone would spend a few minutes completing one of the modules of a new management diagnostic, and then spend the remaining time discussing their results. My request was accepted.

Two of the largest management consulting firms—Ernst & Young and Coopers & Lybrand—were on retainer to provide professional input to the operational audit process. Every audit team had to present their findings to a panel of consultants on an ongoing basis. The panel consisted of three senior consultants from each of the two firms. When we presented our first draft report, the review panel focused on the way we'd written the findings. They gave us constructive criticism on our writing style and taught us how to write an audit report; the wording had to be clear and precise. These six panel members were going to join the audit department for the workshop.

During the workshop, I distributed the leadership Value Analysis Profile. "You'll notice," I said, "that there is a statement on the left-hand side of the page and a statement on the right-hand side. The two statements are separated by a scale ranging from zero to nine. There are two sets of numbers, one below the other. Please read the statements on both sides of the scale and circle the number in the first line that best represents the current situation at your place of work, Hydro employees for Hydro and consultants for your place of work. Circle the number in the second line to indicate what you would ideally like to see. When you complete the scoring, I'd like you to discuss the size of the gap, if any, between the actual and the ideal. Also, I'd like you to talk about the impact this gap may have on your own productivity, and discuss what would have to be done to close the gap. In about twenty-five minutes, I'll ask you to share some of your findings and thoughts."

To my surprise, there was no resistance to the exercise and a lively discussion ensued. At the end of the twenty-five minutes, they reported on their findings and gave a summary of their discussion. Everyone from Hydro as well as from the consultants reported gaps between the real and the ideal. The gaps ranged from as little as three points to as many as eight. There was unanimous agreement that they would like to see changes in the behaviors of their immediate supervisors. When I pressed them, especially the consultants, about the possibility of using this type of diagnostic on a broader scale, there was reluctance.

"In our consulting practice, we interview senior managers and some middle managers to get a sense of the issues that are facing their organizations. We use a very similar approach that you use when you conduct your operational audit. We don't use questionnaires."

"Do you see any value in assessing the values of an organization?" I asked.

One of the consultants responded, "The exercise on values was interesting and the discussion was great. But I can't see how this diagnostic can help in a management consulting practice. Values just aren't considered. We look at operational issues like process effectiveness, efficiency, and economy. We also do an assessment of technology. We make recommendations to solve the problems that we identity."

I shared this feedback with Graham. I told him we'd had enough time to complete one Value Analysis Profile at the audit management workshop. I'd chosen the leadership profile. Everyone seemed to enjoy completing the profile and the discussion that followed was amazing. They identified many behaviors they would like to see changed in their supervisors. For example, most said they would like to be able to trust their immediate supervisor a lot more than they do. They wished that their supervisors shared information more openly. They wanted ongoing feedback on their work, rather than waiting once a year for their "dreaded" performance review. They agreed that they would be more productive if there were changes in the behaviors of their supervisors. When I asked the consultants if this kind of diagnostic tool would be useful in their practice, however, I received a definitive no.

I looked at Graham and he just shook his head in disbelief. The business market was not prepared to consider this type of measurement in the early 1980s. We had little choice but to shelve the Value Analysis Profiles and move on.

Graham asked me to join the board of directors of the KBC. The whole operation was financed by donations from the friends of the KBC and by revenues generated by the various programs. Graham had a knack for attracting business people who either shared a strong faith or had concerns about values in the workplace. He was able to draw these individuals out from behind the glass and steel corporate walls. We organized a number of programs, but the most successful were the many lunch groups that met around downtown Toronto. These groups would meet for discussion and support. They talked about their faith and how they could apply it in the workplace.

I joined a group that Graham had set up for senior managers and professionals. We met in the boardroom in the offices of a legal firm. The senior partner, Jacques, shared Graham's vision of an ethical workplace where there was no separation between the values of our faith and the values of business. We met every two weeks. The meetings were attended by one or two other lawyers, and the rest were typically business people at a president or vice-president level. These were golden times. There were usually twelve at each meeting. We had interesting discussions and we all became lifelong friends. We talked about business problems and helped each other deal with these problems from an ethical perspective based on Christian values. We supported each other and prayed together.

The seeds for a center for business ethics were planted at this time. The Centre for Ethics and Corporate Policy was established by the King Bay Chaplaincy in 1986, and spun off completely from the KBC in 1988 as the Canadian Centre for Ethics and Corporate Policy. This was the first center for business ethics to be established in the world and took two years to create. Although Graham and members of the KBC were the driving force behind the creation of the Centre, Graham recognized at the start that the Centre had to be completely inclusive. Members from all religious groups participated in the formative meetings to shape the Centre, its mandate, and constitution. Members of the first board represented most of the major world religions.

Initially, few paid much attention to the Canadian Centre for Ethics and Corporate Policy. Today, however, the Centre, known as EthicsCentre.ca, is thriving and making a valuable contribution to the business community through its programs and conferences. Ethics and values are part of the regular curriculum at most business schools. Graham was a true visionary and was always ahead of his time.

The intensity of my involvement with the King Bay Chaplaincy was at its highest during my time in audit, but that time was rapidly coming to an end. The typical duration in audit spanned twelve to eighteen months. At the end of this period we either returned to our old jobs or we had to find another management position in the corporation.

Company Politics

I had no desire to return to my old position. It would have been too difficult after experiencing new growth and learning so much about other parts of Ontario Hydro. In the eighteen months with audit, I was part of four audit teams and we completed audits of the Pickering Nuclear Station, research division, the conservation division, and thermal division.

Ontario Hydro was going through major restructuring in 1982–83. A new marketing division was created, and with it, a new position of manager, marketing planning. After a lengthy internal competition for this new position, I was informed that I was the successful candidate and started my new job in September 1983.

The marketing planning department was initially made up of people from the previous market research department. The manager, Len, and his staff reported to me. I was given a budget to hire three additional managers and sufficient staff for three new marketing planning sections: residential, commercial, and industrial market sectors. I also had a three million dollar budget for market research.

At around 10:00 in the morning on my first day in the new position, a portly man burst into my office. He looked like Newman on *Seinfeld*, except he had a black goatee. "My name is Frank," he said. "I'm the union steward representing the unionized workers in the market research section of your department. I want you to know that there are eight grievances that have been languishing for several months, and I want action. Now that the new department has been created and a new manager installed, I want these grievances to be resolved immediately. If not, I'll file a formal complaint against you and your department."

I was completed bowled over. What just hit me? "I wasn't aware of any grievances but I'll look into them," I said. I didn't know what else to say.

"You better. I'll expect to hear from you in a week." Frank left as quickly as he'd come in.

My new job presented a whole set of problems and challenges that I'd never encountered before. I'd inherited an old department, including its manager, who was a twenty-five-year Hydro veteran, and eight union grievances. I was to amalgamate this old department with three new managers that I would have to hire. These managers, in turn, had to hire additional staff for each of their sections. While this was taking place, there was pressure to develop a new marketing plan and marketing strategy for a $25 billion company. I didn't think things could get worse, but they did.

A couple of months after I started, a new director was hired for a sister division, the marketing programs division. Dave was recruited from Camco, one of the largest manufacturers of appliances at that time. The new director was a lanky man, who was well over six foot two, in his late forties or early fifties. From the day of his arrival, Dave's gun sights were set in my direction. It became apparent that he had expected to have the marketing research *and* marketing planning function reporting to him. This was not an unreasonable expectation. At Camco, as vice president of marketing and sales, he'd had full responsibility for creating and implementing marketing programs. However, at Ontario Hydro, which at the time had a $25 million annual marketing budget, these functions were split into two separate divisions: marketing services and marketing programs. Dave was responsible for developing and delivering marketing programs, but my marketing planning function was not under his span of control. The marketing planning function was allocated in the marketing services division, which was responsible for marketing planning, marketing training, setting electricity rates at the wholesale level, and approving the retail electricity rates distributed by approximately two hundred municipal electrical utilities. This division also set the export energy rates.

From an organizational perspective, one could argue that the new marketing planning department could have been located in either the marketing services or the marketing program division. Marketing services was responsible for all aspects of energy rates. One of the functions of the old market research department was to monitor electricity pricing across North America, in Europe, and other countries around the world. Ontario Hydro was in a unique position. It generated electricity, distributed bulk electricity to municipal utilities, regulated the pricing of electricity at the retail level, set prices at the wholesale level, and competed directly against other forms of energy in the industrial and rural markets. In fact, Ontario Hydro was the largest retail utility in the province, because it provided electricity to all the rural areas of the province that were outside the boundaries of a local municipal utility. From a strictly business perspective, it was not unreasonable for Dave to think that the marketing planning function should be located within his marketing program division.

It soon became apparent that the new director could not live with this artificial split. He wanted to have my department under his control. The assaults began on my department and on me personally. Dave was a seasoned corporate gamesman and a ruthless corporate politician. My life became a living hell. First, I was not a political animal. I liked looking after my staff and getting the job done. Second, I was at a disadvantage because Dave was one level above me. He was a director and I was a manager. I was totally outclassed.

I was besieged from all sides. Union issues had to be resolved. New staff had to be hired. I was given six months by my vice president to prepare a marketing plan for Hydro. Then I discovered that Len, the manager of the old market research department, was sabotaging everything that I was personally attempting to accomplish, and he was sabotaging the department's effort to get launched and do the necessary work to complete our marketing plan. For example, Len received monthly publications that contained raw data on energy usage by the different forms of energy. These statistics were broken down by market segments. I asked Len to prepare a summary report as soon as he received this data, so that we could use it as input into the market segmentation analysis that was being done for the marketing plan. The reports were never forthcoming. Len was slippery. He always had an airtight story to justify why the reports were not completed.

It didn't take long for me to discover that Len had expected to become the manager of this new, vastly expanded marketing planning department. The new position was a higher-level manager's position with a higher salary and higher level of approvals. Len had been with Hydro for twenty-five years. He had been manager of the market research department for over ten years. The new manager position was rightfully his. He was entitled to it. He didn't like and could not accept the fact that he didn't get the job.

The new position turned out to be much worse than my job in property. At least in property I was able to get control of the problems and work systemically to resolve them. In marketing, I felt I had very little control. I was working twelve-hour days in an effort to keep my head above water. More often than not, I felt like I was drowning. Luckily, I managed to maintain my eight-mile run each morning before work.

Looking back on those days, I don't consider them to be my finest. I won the battles with the unions and I fired Len, who kept sabotaging the department. It's not easy to terminate a twenty-five-year veteran from a public utility. I would meet with him on a weekly basis to set objectives and expectations that would be summarized in a memo. We would also review the previous week's objectives. Usually those objectives had not been met and the work not completed. I documented everything and, after six months, I had a three-ring binder filled with all of the memos documenting Len's non-performance. One morning, I called him into my office.

"I'm going to come right to the point, Len," I said. "I'm very sorry it's come to this, but I asked you to come in this morning to tell you I'll be terminating your appointment with Ontario Hydro."

There was a minute of silence and than Len went on the offensive. "You can't terminate me. I've been with Hydro for almost twenty-six years. You don't have the authority to do this."

"I do have the authority," I said. "Human resources will prepare a fair termination package that will take your twenty-five years of service into consideration."

When he threatened to take me to court, I showed Len the three-ring binder with the record of all our discussions and his non-performance. He walked out and that was the end of Len.

I did a small reorganization. I allocated Len's ten staff under one of the three sections. It wasn't a perfect solution but it worked. The three managers that I hired were a joy to work with. They were bright, they were committed, and we had a common enemy in Dave the director and the escalating turf war. It's amazing how a common enemy can create a strong, cohesive group.

Turf War

Where was my boss, the director of the marketing services division, in this turf war? To begin with, Hedley Palmer was a brilliant mathematician. In many respects, he was the ideal person for his position from the perspective of the rates side of the business. He understood the complexity of electricity pricing better than anyone anywhere. A man in his early sixties, Hedley was a very approachable, fatherly figure. He had been with Hydro for over thirty years. He was highly respected because of his keen intellect and overall manner. In a sense, he was the elder statesman of the entire marketing division. In addition, his influence had a long reach—into the president's office and throughout the two hundred municipal electrical utilities and electrical associations. Everyone knew Hedley Palmer. It was no wonder that he had the ear of the vice president of marketing. Hedley was also a master strategist, politician, and corporate gamesman.

Looking back, I can see that Hedley was in complete control. He let Dave play his games, attacking my department and attacking me personally. Dave did this in many ways. I had set up an inter-divisional working group that was made up of Dave's three program managers (one for each of the three markets—residential, commercial, and industrial), to work together on the market segmentation, market research priorities, SWOT analysis, etc. Dave always found fault with something: It took too long to complete the segmentation. Why was I wasting his managers' time on this committee? His managers weren't involved enough in developing the marketing plan. His managers weren't getting enough information for their programs. They couldn't develop their marketing programs until the marketing plan was completed. They didn't need the marketing plan—they'd go ahead and develop their programs based on what they saw as being important.

We developed the first draft of the marketing plan, and Dave was all over it. Our overriding positioning strategy was that electricity is a unique form of energy that should be promoted where it has a unique advantage over other forms of energy. The reasoning was that electricity should be considered a strategic energy because there were uses for which only electricity was suitable. Over 50 percent of electrical energy in Ontario was generated by nuclear plants, and that number was going to rise over time. It was a manufactured product. Therefore, Hydro should seek and promote innovative technologies that could only be operated using electricity. For example, in health care, there was new imaging equipment. In the industrial sector, plasma technology in steel manufacturing provided superior benefits over traditional methods of production. In the commercial sector, there were new energy management systems designed to increase the energy efficiency of both heating and air conditioning. In the residential sector, there were smart meters and demand management methods that could flatten load curves.

These were technologies that could only use electricity, and they provided significant benefits to the consumer. The first draft of the marketing plan was based on the premise that electricity is a manufactured product and, as a manufactured product, it should not be wasted on low-end uses, such as space heating or water heating, where other forms of energy could do the job just as well.

Dave did not accept this premise. He wanted to promote the use of electrical energy everywhere. He was a marketer, after all. Before he came to Hydro, Dave was the sales and marketing vice president for an appliance manufacturer. He'd arrived at Hydro with a sales focus. Now he wanted to sell electricity. His residential group developed a marketing program called "Stamp out cold feet with electric heat." This program was clearly positioned to compete against natural gas and oil in the residential space-heating market. I couldn't believe that Hydro was willing to waste electricity on this kind of use. Our first marketing plan was completely rejected.

Hedley remained silent and the vice president of marketing supported Dave. His rationale was that, as the new nuclear reactors were coming on stream, we had excess power that had to be sold. He didn't give much thought to meeting future demand. We had too much now, and so we needed an aggressive positioning strategy for electricity. My department had to comply. We had no allies. We worked twelve-hour days and weekends to complete a marketing plan that aggressively promoted electrical energy in all three market segments. A great deal of pressure was exerted on us because the first marketing plan had been rejected, and now we were behind schedule. A new marketing plan based on an aggressive strategy was presented three months later and approved.

Interestingly, six months after the "Stamp out your feet with electric heat" campaign was launched, the minister of energy saw a large billboard displaying this slogan and a picture of cute feet running across a floor. Apparently the minister went ballistic. He called Hydro's chairman and asked him why Hydro was promoting electricity for residential space heating. The program was shelved the next day. The billboards were a small part of a much larger campaign that included regular TV spots, print advertising, and promotion. It was a million-dollar program. The money was wasted and damage was done internally at Hydro. This aggressive marketing stance remained for a few more years in more subtle forms. Finally, greater external pressures forced Hydro to change its course.

Dave never got his wish. He never got direct control of my department, but he fought hard and he succeeded in directing the course of our work. Hedley seemed content to sit and watch the events unfold. My department was still under his span of control. His empire was still in tact. He maintained the head count of his division, which was considerably larger than Dave's, and Hedley had the biggest budget in the marketing division. From an internal corporate perspective, Hedley was still king of the hill. He was above the fighting. He didn't get his hands dirty, and he did a masterful job of protecting his turf.

Hedley and I traveled on business on many occasions. He loved good food and was a wine connoisseur. We always ate in the best restaurants and had the finest cabernets or chardonnays. These were the heady days of Hydro, with massive budgets and a gold-plated culture. I really liked Hedley because he was approachable and very easy to talk to. I spoke with him very openly and shared many of the issues and problems facing me that needed to be resolved.

But I recall my first annual performance review. It was okay, except Hedley told me he couldn't recommend me for an upgrade in salary (I was at the 96 percent level of my salary range) because of the unresolved problems I still needed to fix. He cited all of the things I had shared with him in confidence, over dinner and a bottle of wine, as examples of why he thought I should remain at the same salary level. I realized that Hedley was not to be trusted. I enjoyed many more dinners and bottles of fine wine in his company, but I clammed up. I never again confided in him or asked his advice or opinion on sensitive internal matters.

Although the pressure of my job was relentless, the running kept my blood pressure in check. My strong faith kept me spiritually rooted. The combination of these two factors allowed me to cope with the stress and maintain my focus. However, I began to lose some motivation and was no longer enjoying my work. I wasn't happy and I developed a healthy cynicism about Ontario Hydro. The irony in all of this is that Hydro was very good to me.

During my three years in marketing, Hedley sent me on executive management training programs at the Graduate School of Business at Columbia University, in New York, and a year later at the Wharton School of Business in Philadelphia. These two opportunities gave me incredible times of learning and allowed me to meet exceptional executives from all over North America. At both of these universities we were put into study teams. At Wharton my study team comprised managers from Apple Computer, Hewlett Packard, IBM, and other Fortune 100 companies. I learned a great deal from the course work and from my study partners.

In 1986 and the beginning of 1987, I was approached by six companies, either directly or through headhunters. I left Ontario Hydro in April of 1987 and joined Avstar Aerospace Corporation as vice president, marketing and sales. This was a small high-tech company that provided life support systems for the U.S. space program. I was hired because Avstar was planning to migrate the breathing technologies they'd developed for the space program into commercial uses in the health and safety field—specifically, providing their breathing apparatus for fire-fighting and mine rescue.

Two years later, I was offered a position as director of marketing at a national energy company (the fictional Northern Energy). I accepted the job because the energy business was in my veins. In many respects, this was a dream job, because now I had the control that I didn't have at Ontario Hydro. I had the complete span of control over market research, marketing planning, and marketing programs. I had everything that Dave had wanted at Hydro. I was in a position to compete against electricity right across the continent.

Section III

The Boss from Hell

The Boss from Hell

You can build a throne with bayonets, but you can't sit on it for long.

—Boris Yeltsin

In August 1990, I was approached by Don Edwin, president of Northern Energy. He asked me to work for him because of my marketing experience at Ontario Hydro and the fact that I had worked in the Maritimes for five years. His company was experiencing stiff competition in the Maritimes, and he wanted someone to prepare a marketing plan for the four Atlantic provinces. He thought that my knowledge of the Maritimes, combined with my knowledge of the energy sector, was the perfect skill set needed to develop the Maritime marketing plan.

I started working as a special consultant to the president at the beginning of September 1990. Northern Energy was a national energy company with over 125 branch offices coast to coast in Canada and the United States. Over the next four months, I spent 60 percent of my time traveling throughout the four Atlantic provinces, meeting with all of the branch managers, and meeting regularly with the regional director, Tom McAllister. I gathered information and statistics on the energy markets. This allowed me to develop a detailed profile of the competition, customer usage patterns, trends, forecasts, and our strengths and weaknesses.

During the week between Christmas and New Year's, I was working in my cubicle back in Toronto when my phone rang. It was the president. He asked me to come to his office.

"Michael, please come in. Sit down," he said. Andy, the vice president of marketing, was sitting in a chair to the left of the president's desk.

It was an enormous office. Although I had been in Don's office at previous meetings, I was always overwhelmed by its size. It looked to be about forty-five feet wide and twenty-five feet deep. The wall behind Don was clad in floor-to-ceiling glass that ran the full width of the office. It faced west. In the winter months, his vertical blinds were typically open until mid to late afternoon, when the sun finally came around the building. The office was not only large, but it was beautifully furnished in dark-stained cherry wood. The reddish hue from the highly polished wood bled through the dark stain. I could see my own reflection in the tabletop. Don sat behind his enormous desk, which was at least 50 percent larger than a standard four-by-eight sheet of plywood. There was nothing on the desk except a red folder. The telephone was on the credenza behind Don's desk. He sat in a black leather chair.

"How's the marketing plan coming along?"

"I've completed all of my field work, data gathering, and research," I replied. "I'm now in the process of setting up the structure for the marketing plan, and I'm about to begin the analysis and synthesis phases."

"I've received positive feedback about you and your work from the field," Don said. "I'll get right to the point. The position of director of marketing is vacant. I'd like you to start in this job effective January 2nd."

While Don continued to talk about compensation and perks, I knew I didn't have to think about my reply. I liked the job and I liked all of the people I'd met in the field. I got along well with Andy, who would now be my boss. The base salary was attractive. I would also get a company credit card that I could use at any gas station and at any of the company stores. These stores sold appliances, barbecues, barbecue accessories, and a large variety of kitchen utensils. Plus, I'd be getting a car allowance.

"Now," Don continued, "I'll expect you to complete the marketing plan by March 1st while assuming the full complement of the marketing director's responsibilities."

I knew it would be a lot of work, but I was up to the challenge. In addition to this career opportunity, there was something else to look forward to. In keeping with the image of opulence in Don's office, the head office building came with a state-of-the-art fitness center, located in the basement. The walls were all mirrored from floor to ceiling. There was an excellent sound system and the showers and change room were first class—better than any private club I'd ever been in. The exercise equipment was the best money could buy. Everything was there: treadmills, exercise bicycles, rowing machines, weight-lifting stations and elliptical trainers. All of these machines were equipped with the latest in electronic gadgets. The bicycles and the treadmills had various settings for terrain and a wide array of options for visual scenery. You could monitor your pulse, heartbeat, and blood pressure. Although this is commonplace today, it was all leading-edge stuff back in 1990.

"Yes," I replied. "I'd be happy to accept."

I saw this as an opportunity to run my own marketing organization where I was in charge of the whole marketing process, as well as getting back into the kind of shape I was in back in my Hydro days.

Head office was located on the outskirts of Toronto in a modern five-story building that was completely clad in green tinted glass. The company had moved into their new headquarters a year earlier. The building was magnificent. I had been impressed with the lobby the first time I'd walked into the building for my final interview with the president. All of the initial interviews had been off-site. The lobby took up about one-quarter of the building footprint. In other words, it was grand. It had marble floors. The glass interior walls exposed offices that faced into the atrium. The atrium rose up three floors, forming an elegant open space.

Day One: Monday

I moved into my new office. It was a miniature of the office occupied by Don, except for the furniture—mine had a dark-stained oak desk instead of the cherry wood that Don had. There was a large black leather chair, and behind the chair was a matching credenza. To the right of the desk and credenza was a dark oak wall unit about six feet wide and seven feet high. The opposite wall was clad completely in glass. In the opposite corner to the wall unit, on the window side, was a dark oak, round meeting table with four black leather chairs. The office faced west. I loved this office. It was the nicest one I'd ever had.

I sat down in my chair behind the desk and looked around my office and out the window. I felt like a million dollars. I had the perfect job with the perfect office. All the furniture was placed at right angles, and I decided to do one thing before starting on anything else. I got up and dragged the left side of my desk to a forty-five-degree angle to the wall and my credenza. This way I could still see the entrance to my office out of the corner of my eye, but now I was directly facing my meeting table and the window. I had placed my desk at a forty-five-degree angle in all of my previous offices. This was my signature. A few minutes later Cheryl, my new secretary, walked in.

"I should tell you that Mr. Edwin doesn't like to have any of the furniture moved. Everything has to be the same as you first saw it."

I came back with a short quip. "I guess there's a first time for everything. I'll be the new trendsetter."

"Well … I'm not so sure this is a good idea," said Cheryl. "I really think you should move your desk back to its original position."

"Let me think about it. I'll just leave it for now."

A few minutes later, Andy came by. "Can I come in?"

"By all means, Andy. How are you today?"

"I'm fine. I just came by to see how you were settling in. I see you've made some changes to the office layout. Let me help you put your desk back in its original position."

"Pardon me?"

"There are a few things I need to go over with you that we haven't had a chance to talk about yet. First, Don wants everything to look exactly the way he wants it to look. This whole building is, in some respects, a monument to Don. It's beautiful and everything is perfect down to the last detail. That means that all the offices must be perfect, as he defines it. So, your desk has to go back to its original position. By moving it, you've disrupted the perfect symmetry of your office."

He bent over slightly and grabbed one end of the desk. "Help me out; lift the other end."

I lifted the other end of the desk and we moved it back into its original spot.

Andy continued to speak. "Also, you cannot leave anything on your desk, your credenza, or meeting table at the end of the day. I don't mean just work files, papers, and pens. I mean all the surfaces of all the furniture must be left bare. There cannot be anything lying on any surface at the end of the day—no knickknacks, pictures, nothing. Pictures of family or any other personal mementos are frowned upon, but if you wish, you can have one framed picture. That's it."

I stared at Andy in disbelief.

"Don frequently walks around the building after everyone has left, and he'll admonish anyone who has left the smallest article on their desk. There are no exceptions to these rules. It doesn't matter if you're a VP or a clerk; the rules apply equally to all of us.

"There's one other thing we need to talk about," Andy continued. "When Don made you the offer of employment, he told you about the company credit card and the monthly car allowance. See Peter Schultz, our corporate lawyer, later on this morning. He has some papers for you to sign and he'll make sure you're properly documented so that you can begin receiving your pay. He'll also give you a company credit card. I need to go now. Give me a call if you have any questions."

"Thanks, Andy, I'm sure I'll be calling you. But I've got to ask you one question before you go. I need to know what your priorities are for the marketing function. What are your expectations of me? Is this something you can talk about now or can we arrange a meeting for later in the week?"

"I'm going out of town tomorrow for the rest of the week on a new project. So, in a nutshell, the reason you were hired was to strengthen the marketing function. We intuitively feel that we could be doing more with the current marketing program. Your job will be to move us forward. We can talk some more next week when I get back."

"Thanks, Andy. I'll see you next week. Have a safe trip."

Shortly after Andy left, I became aware of a bit of a commotion outside my office. Then everything became eerily quiet. At first I thought I was being paranoid; I was still trying to internalize my conversation with Andy. But my curiosity got the better of me and I walked out to see Cheryl.

"Is it my imagination or is something going on out here?"

Cheryl's eyes were a little moist. "Don just fired Carol, John, and Darren. Carol worked in PR, just down the hall. John was the manager of accounting, and Darren was an assistant manager in operations."

"Do you know what they did to get themselves fired?"

"Nothing that any of us can tell. Each year, Don fires three people at the beginning of the year. He wants everyone to know that when he threatens you with firing, he'll actually do it. I really liked Carol and I'm going to miss her. She's worked here for the past three years and we became good friends."

I walked back into my office and checked my watch. It was 8:30. I had been here for only half an hour and I had already lost that initial excitement of a new job. What was going on here?

I realized I hadn't even met my staff yet. I had three marketing managers reporting to me, and they each had a couple of marketing specialists. It wasn't a large head office complement. I also had functional responsibility for approximately 125 sales personnel in the field.

I asked Cheryl to come into my office. "Have any meetings been scheduled for me over the next couple of weeks?"

"Yes," she said. "I was planning on going over these with you when you got settled. There's a weekly senior management meeting every Friday morning at 9:00. These meetings are held in the boardroom on the fifth floor beside Mr. Edwin's office. They can run anywhere from two to three hours, depending on what's on the agenda. Also, Mr. Edwin would like to meet with you on Wednesday morning at 9:00 in meeting room A on this floor."

Cheryl was very professional and, at the same time, there was something very restrained about her. She was tall, slightly heavy set, with long, light brown curly hair. She looked like she was in her early thirties. She continued in her soft voice, "On February 10, all of the regional directors are coming in for two days of meetings. You're expected to attend on both days, but I don't know if you'll be asked to make a marketing presentation at this meeting. That's all I've got."

"Thanks, Cheryl. I noticed quite a bit of stuff in my in-basket. Can you tell me a bit about paper flow and time-sensitive material?"

"Yes. The most time-sensitive forms are marketing requisitions that come in from the field and require your signature for approval. Each branch office has a budget to spend on sales promotions and sales events. Although these are in their budgets, you need to approve the expenditures. The rest of the material that comes your way includes the usual inter-office memos, notice of meetings, announcements, etc. If you like, I can scan anything that comes in and give it to you sorted."

"That would be very helpful. Will you put all the requisitions in one pile and sort the rest into two piles—urgent or important, and information?"

"Yes, I think that's best" she said. "In the case of meetings, most meetings are set up by telephone, so I'll keep you apprised as they come in. If a meeting comes along in a memo, I'll note that as well."

"That's great, Cheryl. I really appreciate your help. I've got to go down and see Peter Shultz now. When I get back, I'd like to meet my staff. Can you find a meeting room and ask everyone to meet with me at 10:00? Thanks."

Peter Schultz, the corporate lawyer, was located on the third floor. Even though the door to his office was open, I walked over to his executive assistant and introduced myself. I asked her if Peter could see me. She stood up and walked me to Peter's door.

"Mr. Schultz, this is Mr. Koscec. He's here to see you."

"Come in, Michael. Welcome. I was told you'd be coming by this morning." Peter was about five-foot-eight with a slight build. He had short, sandy hair.

"Hi Peter. I'm pleased to meet you. Andy said that I needed to see you this morning so that I can become official."

"Yup. I've got your credit card and I need you to sign something. It's our standard practice to have all management and professional employees sign a confidentiality agreement. Briefly, this agreement says that you will not disclose any information about our company to anyone, and when you leave the company, you will not work for a competing company for two years after your departure."

I thought this sounded reasonable and I signed the agreement.

When I returned to my office, I went through my in-basket and, before I knew it, it was 10:00 o'clock. Cheryl had booked meeting room B on our floor. When I walked into the room, it seemed that everyone was there. I was impressed. They were very punctual.

"Hi," I said. "My name is Michael Koscec. I'm sorry we're meeting in this conference room. I was planning to come around and meet all of you on a one-on-one basis, but it seems we need to meet sooner rather than later. Could we go around the room and introduce ourselves, and you tell me what your responsibilities are?"

I learned that the three managers were given responsibility for a number of market sectors. They had divided the energy market into twenty segments, and each manager had responsibility for several related sectors. I also learned that there was a marketing "playbook" (my term) for marketing programs. All the marketing programs were catalogued in three-ring binders. These binders listed all the details for every program—a description of the program, a description of how the program was to be implemented, and the start and end dates. For example, every September they ran a 10 percent discount on residential furnaces that started a week after Labor Day and ended on the last Saturday of September. The program consisted of adding a discount flyer to all of their customers' bills. Each month, a series of marketing programs would come out of the playbook. The only thing the marketing staff did was work with their advertising agency to produce new flyers with a new look, updated start and end dates, but the same message. The whole marketing program was mechanical.

Don had told me to improve the marketing function, and this was part of the reason he'd hired me. Andy had mentioned the same thing earlier in the day. Based on what I'd seen and heard today, there was a great deal to do. But I still had no clear idea what Andy and Don had in mind.

Day Two: Tuesday

I divided Tuesday into two halves. The first part of the day was spent understanding the requisitions that needed my approval, and how they tied in to the overall marketing program. There was no marketing plan or advertising plan, so there was no way to assess the value to the company of these local program expenditures. The other half of the day was spent going through seven three-ring binders that contained all of the marketing programs. I also discovered that there was a file containing a monthly report of all energy sales by market sector. There was another report that contained the monthly sales data for all the appliances, etc.

Day Three: Wednesday

I decided to start using the physical fitness center. I arrived at 6:00 AM and started a routine where I'd work out for an hour, shower, and be in my office around 7:30. I had gone off the physical fitness program I'd started at Ontario Hydro. I'd had to stop running because my knees and my back were starting to bother me. In fact, I went to a knee specialist who gave me a choice: Stop running or you'll soon stop walking. As a result, I'd gained twenty-five pounds and was not in the best of shape. This was a perfect opportunity to get back into shape. I couldn't use the treadmill, but, between the stationary bike and the rower, I was able to raise my heartbeat and work up a good sweat. I started out slowly and worked my way up to higher tensions.

The facility was great and I had it all to myself. I even set the sound system to my favorite radio station. I was able to work out, and at the same time, hear the latest news. I didn't push very hard on my first day, but I worked a little harder each day, and, by the beginning of March, I had dropped twenty pounds and reached a level of fitness where my energy was high and I felt like a million dollars again. I was almost at the same level I'd been at Ontario Hydro, when I ran eight miles each morning before work.

The first day of my workout was the day I was meeting with Don at 9:00 AM. He hadn't given me an agenda, and I wasn't sure why he wanted the meeting. I figured he wanted to talk about my work objectives. I was very much aware of the fact that I had put the Maritime marketing plan on the back burner, and that I needed to find a block of time each week to work on it. Don was expecting a completed plan by March 1.

Meeting room A was to the left and down the hall from my office. I arrived at a couple of minutes to nine. Don was already sitting in a chair at the conference table.

"Good morning, Don."

"Good morning, Michael. How are you today?"

"I'm fine, thanks. I really like your physical fitness center."

"It's state of the art. Are you planning on using it?"

"I started this morning. I like working out each morning before work."

"Good," said Don. "Now let's talk marketing. What's happening with the Maritime marketing plan?"

"I've completed the data gathering. Now I'll start on the analysis and the marketing strategy."

"Don't take me into the weeds here. I don't want to hear your life history. When will it be finished?"

"You'll have it March 1st."

"All right. So, what's happening on the marketing front?"

"I met with all the staff to find out what everyone was doing. I've started an assessment of all the marketing program binders."

"That's a pretty good piece of work."

"Well, yes and no. I see a lot of programs, but it's not clear what the overall marketing strategy is. I've been told there's a marketing plan for the Western region. Can I see that?"

"No, you can't," said Don. "I like the Western marketing plan and I don't want you to see it until the Maritime marketing plan is completed. I want to see what you come up with first. I don't want you copying from the Western plan."

"Based on what I've seen so far," I said, choosing my words carefully, "there's a need for an overall marketing plan. Where is the company heading? What are the financial growth targets? What is the implication of these growth targets on market share and sales volume increase? What are customer needs and how can we leverage these needs to grow our business?"

"There you go again, taking me into the weeds, Michael. I want new marketing programs. Do you understand?" Don's voice began to rise. Up to this point, the tone of the conversation had been fairly even. But now, I could see Don losing his patience and getting angry. His face began to turn red.

Although I'd had a fairly good workout earlier, I became aware on the fact that my armpits felt hot and moist. When I'd first walked into the meeting room, Don wasn't wearing his suit jacket, so I'd taken mine off and put it on the back of the chair. As I jotted a few words on my note pad, I could see that the underarm of my shirt was soaked.

"You were hired to strengthen our marketing," said Don, firmly. "I want new marketing programs. If you can't give me new marketing programs, I'll fire you and I'll find someone who can."

He pushed his chair back from the table and rose, saying in a normal tone of voice, "I'll give you two weeks to present a new slate of programs. My secretary Jeannie will call Cheryl with a time and place for our next meeting. I think we're done." He left the room.

I sat in the meeting room for a few minutes. I was stunned. I felt like I'd been run over by a Mack truck. I'd been right. He did want to talk to me about my objectives, but I could never have guessed at the tone of the conversation, nor that my work objectives would be so finite. If I was the director of marketing, I was certainly not being given the scope to do my work. The situation seemed ridiculous. There was a Western marketing plan and I was not allowed to read it. I'd been told to prepare new marketing programs, but what objectives did Don want us to reach with those marketing programs? I was expected to create programs in a vacuum.

I left the meeting room and went straight to Cheryl. It was 9:30.

"Cheryl, can you see if all the marketing staff is available for a meeting first thing tomorrow morning? We'll need a large meeting room. I'll prepare an outline for the meeting by lunch so they'll have time to prepare. Book the meeting from 8:30 to 12:30. If one or two can't make it, ask them to change their plans if they can. But I want all three managers at the meeting. Thanks"

I went back to my office. I needed time to think. A few minutes later, the phone rang.

"This is Sam Schwartz from *Truck and Tractor* magazine," the caller said. "Last year your company advertised in our magazine. We publish twelve issues each year and you ran an advertisement in every issue. I'd like to confirm a renewal of your advertising."

"Well, Sam, I'm new here and I'm not aware of advertising in your magazine. I haven't completed my review of the advertising program and budget for the current year. Let me check into it and I'll get back to you."

"But we're preparing for our next issue. I really need a commitment from you now."

"I'm sorry, Sam. I can't give you a commitment today."

"If not today, what about tomorrow?"

"Not even tomorrow. The best I can do is call you in one week."

Sam didn't sound happy, but he gave me his number.

I started thinking about the meeting with my staff tomorrow. Don wanted a new marketing program. I didn't want to create a program in a total vacuum. We needed more customer data. I had good data for the Maritimes on energy usage by sector. I knew where new and existing opportunities could be found and I knew where we were vulnerable to our main competitor. Did my staff have similar data for the rest of the country? It seemed like a good starting point.

Meeting Agenda

I met with Don Edwin this morning. He wants us to develop a new marketing program, and he wants an outline in two weeks. The purpose of this meeting is twofold:

1. Look at customer and competitor data to determine where the greatest growth opportunities can be found.

2. Brainstorm to develop an outline of a comprehensive marketing program that may or may not include all the market sectors—where can we get the most "bang for the buck?"

In preparation for this meeting, please think about this challenge. Come prepared with your thoughts and ideas. Also, please bring the following:

1. As much customer data as you have at hand.

2. The current advertising plan. (Is there one? I haven't seen it. I found a budget line for adverting and another for promotion, but nothing else.)

3. We need to discuss this year's "marketing playbook" and the programs you're currently working on so that we fully understand the link between what you're doing now and what might be displaced by our new program.

I couldn't think of anything else at the moment, but figured this was enough to get us started. I printed it out and gave it to Cheryl to pass along to everyone.

The rest of the day was spent going through the remaining "playbook," my foot-deep in-box, and several telephone calls from the field offices. During this whole time, I kept thinking about the Maritime marketing plan. I still had a great deal of work to do and I needed to find time to finish it.

Day Four: Thursday

My team was promptly assembled at 8:30. This was so different from Ontario Hydro. I remembered the first meeting with my staff when I started my job as manager of marketing planning. I called for a 9:00 AM meeting. I walked into the conference room a couple of minutes before nine and I was the first to arrive. At nine o'clock, ten out of thirty-five walked in. The other twenty-five waltzed in nonchalantly over the next fifteen minutes, carrying their coffees and, in many cases, their muffins or bagels. I had another meeting scheduled at 10:00, so I was not too pleased.

I was not pleased with what I'd seen so far of Don's management style, but I couldn't help but ask myself if everyone's punctuality was due in part to the fear that Don had instilled in his employees.

"Good morning. Thank you for being here on such short notice. As you've seen from my meeting notice, Don and I met yesterday, and the upshot of the meeting was that Don wants a new marketing program. He didn't say what objectives he wanted to achieve or how this marketing program fits in with the company's business plan. He wants a new marketing program. The way I see it is that this makes our job easier. It means we have a clean slate. So with that in mind, I think the guiding principle for this morning is: What can we do that will make the greatest contribution to the revenue and growth of this company?

"Before we begin," I continued, "does anyone have any thoughts about what I've just said? Can you think of any other guiding principles we should embrace for the task at hand? Do you have any questions?"

Doug Westling spoke first. He was the manager responsible for the marketing programs for the residential sector

"I think our task is clear," said Doug. "Those of us who've been around for a while know that there isn't much discussion. If Don wants something, we have to give him something. Now, in terms of your guiding principle, Michael, if you look at the stats, you'll see that the residential sector is the largest. This sector will give you the biggest bang for the buck and this sector has the highest profile."

"Doug is right about the residential sector being the biggest," said Allan, manager of transportation. "However, the transportation sector has a high profile and it offers the greatest growth opportunities. It's also one of Don's pet sectors. If we leave out transportation, we may be running the risk of falling out of favor with Don even before we've started."

Doug jumped in again. "You can't dismiss our stores, Michael, because they overlap with all of the sectors. You should call Craig Lofthouse. Craig is vice president of merchandising, and he's responsible for all aspects of merchandising in our company stores. This means he's responsible for the "look" of the stores, the way merchandising is displayed, and for the sales of all appliances, barbecues, and accessories. More importantly, Craig is the only vice president with profit-and-loss accountability. All of the stores have their own balance sheets, and these are consolidated at Craig's level. Because of his P&L accountability, and because the stores represent the face of the company, Craig probably has the highest profile of any of the vice presidents. He's a big shot around here, so you need to contact him A.S.A.P."

"I really appreciate this, Doug. Things are happening so fast I haven't had a chance to find out about all the key players."

Dominic, the marketing specialist in industrial, followed. "I believe we need to stay focused. Although the industrial sector is not that far behind the residential in revenues, and—Paul, as manager for industrial, please correct me if I'm wrong—this sector has far fewer customer numbers, and it doesn't have the high profile or cachet shared by the residential and automotive sectors and the stores. Don loves the stores and he loves the transportation sector, but the transportation sector is politically important and is high profile because of the environmental connection."

"Are we all in agreement that we should focus on the residential and automotive sectors and the stores," I asked, "and leave out the commercial and industrial sectors at this time?" I looked around the room and there was a general nodding of heads.

"I've completed reviewing the marketing playbook, as I like to refer to it, and did a quick scan of the kinds of advertising that we do. The conclusion I came up with is that every region seems to be doing similar things at the same time. They all run local ads that are straight out of the playbook, and they're following the playbook from a timing perspective. So they push stoves and fridges one month, washers and dryers the next, barbecues in the early spring, and so on.

"The problem I see with this approach is that it's one-dimensional. Here are the 10 percent discounts that we offer this month. That's what the advertising says and what the tent cards show in the stores. But the approach lacks drama, sizzle, heart, and excitement. I think we need to create a marketing program that's three- or four-dimensional and that's filled with excitement and drama."

"I agree about bringing in some excitement," Allan said, "but what do you mean by three- or four-dimensional?"

"I mean we need to use all the tools in our marketing kit bag. We need to involve our manufacturers, trades, and customers. Let's develop an integrated marketing program that involves all of these stakeholders and that's focused on achieving a set of common marketing objectives."

I looked around the room and saw that everyone was listening intently. I continued, "For example, instead of our typical 10 percent offer, let's develop an outline of a comprehensive marketing program and share it with our manufacturers to see what kind of a price break we can get from them, so we can offer, say, a 25 or 30 percent discount instead of the usual 10 percent. Let's add a customer referral program and trades referral program. Do you see what I mean?"

"There's all kinds of other things we can do as well," offered Joe Kirsch, the marketing specialist for residential. He was a recent recruit from American Express. I got the feeling he was very bright; he'd come to the company with solid retail marketing experience. "We can run a contest for current customers who provide the most number of referrals. We can offer incentives to contractors who recommend our products, and offer incentives for each new installation."

I could feel the excitement growing in the room and, buoyed by this new energy, I got up off my chair and began to pace enthusiastically. "Absolutely," I said. "Now you're cooking. Don't forget the four-step principle of successful sales and marketing: first, create awareness, then interest, then desire, and then close the sale. Your marketing program has a logical sequence and timing. I think I can leave you to develop an outline. And please take a stab at setting some sales objectives.

"Today is Thursday," I said. "I haven't heard from Don yet about our next meeting. He said two weeks. So, we have all of next week and probably Monday of the week after. I'd like to suggest that we meet next Wednesday to review what you've come up with. In the meantime, please feel free to talk to me if you have any questions or if you want to bounce ideas off me. I'll let you know as soon as I hear from Don about the next meeting. I'd like several of you to accompany me to this meeting. This is a team effort and you should all get credit for your work."

I had a really good feeling after the meeting. Even though I didn't like the way the program direction was forced on us, I felt a very positive energy in the group. When I returned to my office, I looked in the company directory and called Craig Lofthouse. To my surprise, Craig picked up the phone. "Hi, Michael. Welcome on board. What can I do for you?"

"Hi, Craig. I'll get right to the point. I met with Don yesterday, and he asked me to prepare an outline for a new marketing program and present it to him in two weeks. I had a preliminary meeting with my staff, and it was obvious that anything we put together will have to involve your division, because the stores play such a key role in any marketing push. What's the best way to involve you and your division in developing the new program? We have some ideas already, and the stores play an integral role."

"Actually," said Craig, "Don's already told me that he asked you to do this." I didn't know it at the time, but Craig's office was on the fifth floor next to Don's. "I'll talk with David Booker. He's my director of store merchandising and the right person to represent the stores' interests. I'll ask him to call you, and you can tee everything up with him. I don't need to get personally involved. David will keep me informed, and if I have any concerns I'll let you know."

"Thanks, Craig. I appreciate your cooperation."

"No problem. I'm glad I can help. I look forward to seeing what you guys come up with." Craig hung up the phone. Although he was helpful and cooperative, Craig's voice and manner were formal. He was all business.

Day Five: Friday

I was in the fitness center at 6:00 AM. I loved starting my day with a good workout. I was in my office by 7:30 and began to read the agenda for my first senior managers meeting. There were five vice presidents and five directors who regularly attended these meetings. There were at least another eight directors that I knew of from the organization chart, but they weren't included. I was looking forward to this meeting because it would give me a chance to meet some of the directors and vice presidents. I have to admit that I felt a little bit special when I discovered that less than half of the directors were part of the senior management team. Up to now, I had only met one of the directors in operations, whose office was close to mine, and, of course, my VP, Andy. I was hoping that Andy would be back from his trip to Chicago because there was a lot I wanted to discuss with him.

I arrived at the boardroom at 8:45. There was no way I was going to cut it too close. This might also give me a chance to meet the other vice presidents and directors. After seeing Don's office, I shouldn't have been surprised when I walked into the boardroom—but I was. The boardroom was about the same size as Don's office. It also had the identical design as Don's office. The difference was the orientation. While Don's office ran in a north-south orientation, the boardroom was longer, running east-west, and narrower north to south. There was an enormous dark-stained cherry-wood board table. The west wall was completely clad in floor-to-ceiling glass. The south wall was covered in the same cherry wood, floor to ceiling. The north wall was white. There was a long cherry-wood credenza against the wall, and two large abstract paintings hung over the credenza. They were framed in—what else?—dark-stained cherry wood. The boardroom table was surrounded by elegant black leather chairs with cherry-wood accents. There was a door just to the right of the credenza.

There were a few people in the boardroom when I arrived. They occupied the chairs at the far end near the glass-clad wall. I had lots of empty chairs to choose from. I chose the third chair from the end on the south side, against the cherry-wood wall. After I put my note pad at my place I walked over to introduce myself to the others. I was particularly interested in meeting Craig Lofthouse and David Booker. I assumed David would be included in the senior management meeting.

"Hi. I'm Michael Koscec, director of marketing."

"Hi Michael. George Cornell, vice president of commercial."

"That includes auto and transportation as well, doesn't it?"

"Yes, it does."

"I'm glad we met. I haven't had a chance to call you yet, but we need to talk. Can I call you later today?"

"Yes, sure." George was very approachable and friendly. "What would you like to talk about?" he asked.

"Don has asked me to put together a new marketing program," I said, "and my guys felt that the transportation sector should be included. Although I've got a few dedicated resources for transportation marketing, I think it's important for you or some of your staff to be involved with us if we're going to develop something new—something that hasn't been done before."

"That sounds like a good idea. I look forward to your call."

"Thanks, George." I moved toward the next group and introduced myself.

As I had been talking to George, a group of men had walked in, Andy among them, so I walked back to my chair. Andy sat down immediately to my right. The chair to my left was now taken. It seemed to me that the chairs closest to the west filled up first, and those closest to the entrance at the east end of the room were taken last.

"Hi, Andy. I'm sure glad to see you this morning," I said. "I was hoping you'd be back from Chicago."

"Hi, Michael. How are you settling in?" Andy didn't give me a chance to respond. "No one misses the Friday morning meetings. This is a command performance."

"Can you introduce me to Craig Lofthouse?" I asked. "Is he here yet?"

"Yes, Craig's here. He's sitting at the end of the row across from us on our left." Andy was motioning with his head to a tall, thin, dark-haired man in his late forties or early fifties. Frequently, the image you get from a person's voice over the phone is nowhere near what they look like when you meet them in person. This was a case of the very opposite. Craig looked even more distinguished in person than he sounded on the phone. He was standing near the window end of the boardroom in deep conversation with another man and a woman. The man had wavy blond hair. He was heavy set and I guessed he was in his mid-forties. The woman was tall and slender and had long, straight brown hair. She looked to be in her early forties. Andy checked his watch. It was about eight minutes to nine.

"Come on," he said to me. "Let's go over. We still have a couple of minutes." He stood up and we made our way to the end of the table on our side and walked across to the window.

"Good morning, Craig, Marcia, David. I'm sorry to butt in but I'd like to introduce you to Michael Koscec, our new director of marketing."

"Good morning," they said in unison.

"Good morning. I'm glad to be able to meet you, Craig, after our telephone conversation yesterday."

"Michael, this is David Booker that I spoke to you about. And this is Marcia Hanson, our director of HR."

We exchanged greetings and then I said, "I guess David, you and I will be speaking soon. Will you be in this afternoon?"

"Yes, I will."

"Great. Can I call you around 2:00?"

"Okay," said David. "Talk to you later."

As Andy and I returned to our chairs, he said, "Looks like you've been busy while I've been away."

"I have a lot to tell you, Andy." We sat down.

A few minutes before nine o'clock, the door to the right of the credenza opened. Don walked in with an assistant, Teresa, who was there to take minutes of the meeting. This door was Don's private entrance into the boardroom, directly from his office. He walked to the east end of the boardroom table and sat down. I don't know if I was paranoid, but I immediately understood why everyone sat at the west end of the room—to be as far away from Don as possible. I was soon going to find out that I wasn't being paranoid.

"Good morning, everyone. Who knows what the spot price of oil, natural gas, and propane was this morning? Anyone?"

Jack Burrows spoke up. "Oil was trading at $30.60 on the spot market this morning. I don't know what the others were trading at."

"Okay. Anyone else? No, no one? What's the matter with you guys?" Don said in a slightly raised voice. "You're in the energy business. You should know this stuff. It should be at your fingertips at all times. Especially you, Michael. You're the director of marketing, aren't you?" I felt like I had just been punched in the stomach. I was totally embarrassed.

Don proceeded to tell us what the spot prices were for all the energies. He had a very self-satisfied look on his face.

"Although you've just met him as the guy who doesn't know his energy pricing, I'd like to start the meeting this morning by introducing our new director of marketing, Michael Koscec. Welcome, Michael. I'm expecting great things from you. Now, let's get moving. We have a full agenda this morning and a number of action items from last week's agenda.

"Let's start with operations. Jack, you presented us with your tank inspection plan last week. We made a number of changes to your plan as it related to your proposed schedule. Give us an update on the implementation."

"We made the changes to the schedule as per last week's meeting," Jack began, "and I sent out the new inspection schedule to all of the branch managers. I asked them to shift the technicians away from their current work program to facilitate the new inspection schedule."

Don rose from his chair, his faced a bright tomato red. He slammed his fist on the boardroom table with a loud thud and began to yell at Jack at the top of his voice. "You stupid son of a bitch! You dung-loving, manure-sucking pig! No, that's not accurate! A pig has more brains than you! Who the hell gave you the authority to tell the branch managers to reallocate the resources away from their current work program so that they can meet the new inspection schedule? That was the whole point of changing the timing on the schedule! I wanted them to integrate the new inspection schedule into their existing work program, damn it! How the hell do you propose to fix this fucking screw-up?" Don sat down.

Poor Jack was pale. "I'll get in touch with them and tell them that they have to integrate the new inspection schedule into their current work program."

"And when the hell are you going to do that?"

"As soon as this meeting is over," said Jack.

"Okay. Let's keeping moving forward."

The rest of the meeting was a bit quieter. Don didn't slam his fist on the boardroom table again or get up from his chair to yell. However, he did raise his voice several times and threatened a vice president and a director with firing if they didn't do what he told them to do.

The meeting ended at 11:00. Don stood up and walked to the door leading into his office. His assistant walked out the main boardroom doors, which opened into a reception area. The reception area housed two workstations—one for Don's executive assistant, Mary, and another for Teresa. There was a beige sofa, a coffee table, and two chairs. The decor matched Don's office and the boardroom.

As we were filing out of the boardroom and into this reception area, I turned to Andy and asked, "When can I update you on what happened this week while you were in Chicago?"

"Come to my office at noon. Bring a sandwich and we'll talk while we eat."

"Thanks, Andy. I'll see you at noon."

As I walked past Cheryl, she asked, "How was your first senior managers meeting?"

"I'm not sure how to describe it. I don't know what to say, except that I've never experienced anything like it before."

I walked into my office a little stunned. I felt like quitting, but that wasn't a realistic option. My oldest daughter was attending grade ten at Bishop Strachan, a private girls' school, and my youngest daughter was still in grade school. My wife was working on her master's degree in divinity and we had a mortgage on our house. There was no way I could quit. I was locked in. To make things worse, there was a serious recession in North America at the beginning of the nineties. In fact, it was a worldwide recession that was precipitated by the downturn of the Japanese economy. The recession lasted until about 1996, which saw the start of the dot.com boom. There weren't a lot of jobs around, especially at a director or vice president level.

As I picked up a sandwich at the restaurant on the ground floor of our office building, I remembered how special I'd felt when I realized that less than half of the directors were included in the senior management team. After this morning's meeting, I knew that being excluded was a blessing in disguise and wondered if the excluded directors knew how lucky they were.

Andy's office was next to mine. His door was open, so I peered in as I knocked. Andy was sitting at his desk.

"Come on in, Michael." He motioned with his hand toward his conference table. There was a white paper bag on the table and a cup of coffee. Every floor had a small kitchenette with a table, four chairs, and a professional-style coffee maker. Fresh brewed coffee was always available.

Andy's office was identical to mine and, to my surprise, it was not any bigger. I sat down at the table and Andy joined me.

"I hate to ask you, but are all the meetings like the one this morning?"

"Yeah, pretty much the same. Don has a short temper. No one knows who'll be picked on at these meetings. Usually, there's only one major outburst and one real victim per meeting. He spreads his wrath fairly evenly and doesn't play any favorites."

I was surprised at Andy's nonchalant reply. He talked in a low, even, matter-of-fact tone. No annoyance. No anger. No amazement. In fact, no emotion at all. This is just the way it is—total acceptance. Several weeks later, after I had witnessed several more outbursts at senior management meetings, I would reflect on our conversation and understand his response. It's just the way it was, and there was nothing anyone could do about it, except leave. But we were all in the same boat. We had families, mortgages, and debt, some more than others. We had no choice but to put up and shut up.

"How do you cope?" I asked him.

"First of all, I tread very softly and try not to make mistakes. I keep as low a profile as I possibly can. I've been around long enough to know what sends Don off the deep end, but even I'm not immune from getting strips torn off my back.

I take my lumps and move forward. And I always remind myself how much money I make. That somehow makes it all bearable and helps me get by."

"I appreciate your candor, Andy." I took a bite of my sandwich while Andy unpacked his lunch. "The main reason I wanted to see you," I said, "was to tell you what transpired this week. You may or may not have heard that Don asked me for a new marketing program. He didn't mention any objectives or anything he wanted to see accomplished. He just said he wanted a new program."

"Yes, I heard, and I was going to talk to you about that. How's it going?"

"Well, I assembled the team to determine which marketing sectors we should focus on and to rough out an overall concept. That's done, and now they're putting some flesh on the concept. Do you want to hear what we came up with?"

"Yes, absolutely. But before you start, I should mention that Don frequently asks for something in general, just to see what you come up with. I know you're running blind, but you'll have to take a stab at the objectives and put it all together. I want to get involved in this with you, and I think I'll be able to help. I'm not expecting to travel for a while."

"Oh, that reminds me. How was your trip to Chicago?"

"My trip was fine. I can't really talk about it right now."

I shared the outline of the marketing program with Andy.

"I like what I'm hearing" he said. "You and the team have chosen the right sectors to focus on, and the comprehensive, multifaceted approach is great. We've never had anything like it before."

Andy's reaction meant a lot to me. "Because it involves the automotive sector," I said, "we'll need to meet with George sooner rather than later. How do you want to handle that? Do you want to speak to George, or are you okay if I talk to him?

No, you go right ahead. Bring him into the picture, because it'll involve his resources. You'll have a better handle on this anyway, so there's no point in getting me involved at this early stage.

"You seem to be on the right track," he went on. "I think you have to look at your situation using this analogy: Picture a running train. The train cannot keep running without an engineer to run it. You are the engineer and you have to keep the train (i.e., the business of sales and marketing) running. However, change is needed. You've been hired to bring in change. Specifically, I think Don recognizes that we could be doing better with our marketing and sales. So you were brought in to help us move to the next level. But you still have to keep the marketing enterprise running. It's a balancing act. You need to do both."

"This is really helpful, Andy. And that leads me to another question. How were the 'marketing playbooks,' as I like to call them, developed?"

"They were developed over a number of years by many different individuals in your position. You have to appreciate the fact that Don is a chartered accountant. Therefore, he thinks like a chartered accountant. Everything is ordered and logical. Everything is black and white. Everything has a place and a time. The marketing playbook, as you call it, is essentially a creation of Don's."

"So, it's paint-by-numbers marketing," I said. "You don't have to think or be creative, just pull the appropriate file for January or February, or whatever month it happens to be and, presto, you're away to the races."

"Exactly. But you have to understand where we came from. We came from nothing. There was nothing. This company was built by buying up other energy companies across the country. Finally, when we had a coast-to-coast presence—by the way, this happened before I arrived—there needed to be a consolidation of the marketing and sales functions. The marketing playbooks were the result. That's what we have and I guess it's served us reasonably well. But the energy market has become very competitive over the past few years, and we've lost some market share in some parts of the country. We're feeling vulnerable, even though we're still one of the biggest players. So, the pressure is on us, and, I guess in some respects, more on you than me because I've got other responsibilities. We need to strengthen our business through marketing."

"Tell me about Don's yelling, bullying, and threats," I said. "Was he always like this, or is this recent, or what?" I was really trying to understand.

"As far as I know, he's always been like this. In fact, I heard a story once. His parents were farmers and Don grew up on a farm. He was on a date with a girl he was trying to impress. They were out riding one Saturday evening and they got off their horses and started to walk. Each had their horse's lead in their hand when the girl's horse started acting up. It pulled sideways, got up on its hind legs, and became very difficult for the girl to control. Apparently, Don saw an old two-by-four lying near a fence and, holding the two-by-four like a baseball bat, he smashed the horse across the nose. I think there's a lesson to be learned from this incident, and it might help to explain his behavior. That's just the person he is."

"Wow, what a story! Pretty scary." I was stunned. "I appreciate everything you're telling me, Andy. It helps me understand the lay of the land a lot better." Andy smiled and nodded his head. "I'll get in touch with George," I continued. "Also, we've scheduled a review meeting for the new marketing program for next Wednesday morning at 8:30. The whole morning has been blocked off. It'll be in the large meeting room B on our floor."

"I'll be there," said Andy.

It was almost 1:30 when I left Andy's office. I was surprised how quickly the time had flown by. I hadn't had much of a sense of Andy when I'd started on Monday because we hadn't had a chance to talk. But after our lunch meeting, I really liked him and developed a great deal of respect for the man. He was bright, a clear thinker, and I liked his low-key manner. I was also feeling a whole lot better about the company after meeting the senior management that morning.

When I got back to my office, Cheryl handed me a memo from Mary, Don's executive assistant. The meeting with Don was arranged for Monday afternoon, in meeting room A on the fifth floor. He wanted me to bring no more than two of my staff. I noticed that he had sent the meeting notice to Andy as well. On the one hand, I was comforted by the thought that Andy would be there. On the other hand, I was concerned that we had less than two weeks to prepare. I picked up the phone and called George Cornell.

George answered immediately and I said, "Hi George. Do you have a minute to talk now?"

"Sure. What can I do for you?"

"As I mentioned to you this morning, Don asked me to create a new marketing program. I met with my staff and we came to the conclusion that we should initially focus on the residential and automobile sectors. So we discussed a concept that my staff is now fleshing out. Are you comfortable having the auto sector included?"

"Yeah, no problem. I think it's a good idea, but I just want my key auto people to be involved in creating the program with your staff."

"Absolutely," I replied. "They have to be a part of the team. I'm sure that my guys have already contacted some of your staff. Don gave us two weeks to come up with something. He's arranged a meeting for next Monday afternoon. That's in one week. We're having a review meeting Wednesday morning at 8:30 in meeting room B on our floor. Can you make it?"

"I can join you later on in the morning. I've got a meeting at 8:30 and it should be over in about an hour."

"Actually, come to think of it," I said, "splitting the meeting into two parts is probably a good idea. We could do the residential first and then the auto. I can't see any crossovers between the two sectors. Or, if there are any, they'll be minor. So, coming later will work. Look forward to seeing you."

It was almost 2:00, so I picked up the phone and called David Booker. David answered the phone on the first ring.

I told him that Don had asked me to prepare a new marketing program and provide an outline in less than two weeks. "Clearly the stores will have an important role in any marketing program that we roll out," I said, "so your involvement will be crucial in anything that's developed. Can you spare some time for my guys next week while they're sketching out the program?"

"I think I can make some time available." David didn't sound enthusiastic. He was the first person so far who didn't seem particularly interested in being part of the team.

Sensing this, I added, "We're having a first draft preliminary meeting Wednesday morning. If you're not able to provide input to the guys prior to Wednesday, can you join the review meeting?"

"I'll see what I can do," he said in a non-committal voice.

I wasn't happy with that last conversation, but all the ducks seemed to be lining up. I needed to set some time aside to restart the Maritime marketing plan, but the urgent part of my in-basket had several documents that I thought I'd better look at.

I worked through the afternoon, and at five, Cheryl peered into my office to say goodbye and to wish me a good weekend. Next thing I knew, it was six-thirty. I decided to leave. I went home exhausted and couldn't believe that I had worked at my new job for only one week.

Week Two

Monday and Tuesday flew by. I had several mini-meetings with each team on both of these days. These were meetings that the teams initiated. They wanted to get my thoughts on some ideas, and asked my advice on a whole range of issues. They were working meetings and I was glad to participate, because I had a lot of ideas I wanted to share with each team. Also, after learning as much as I had about Don, I had definite ideas about the way we should present the program to him. We had to keep the verbiage to a minimum and get right to the facts. We had to talk about objectives, numbers, and no reference to the fact that the program was multi-faceted, involving customers, stakeholders, contractors, and customers. We had to dispense with any introductory remarks. There was no point in telling Don how we chose the sectors for the program. We had to jump right in.

One thing became clear at these meetings. There was no advantage to linking the residential and auto marketing programs. When we realized this, we agreed that on Wednesday we would run through the residential program first, followed by the auto program. We also renamed the auto program the transportation program because it included both cars and trucks. We agreed that each team was welcome to come to the other's presentation. I was impressed with the progress both teams had made.

The first two days of my second week did not go without a hitch. About nine o'clock on Tuesday morning, Don called me.

"I had a call from Sam Schwartz from *Truck and Tractor* magazine, and he told me that you didn't want to advertise in his magazine."

I immediately interrupted. "I told Sam I was new and that I needed a few days to review the advertising program and advertising budget to see how his magazine fit into the marketing mix."

Now it was Don's turn to interrupt, before I could tell him that I'd promised to call Sam this week. "I don't give a shit if you haven't had time to review the advertising program. Just renew the ad."

I protested. "But, Don, what is the advertising supporting? I've asked for a copy of the magazine and for a copy of the ad, and I haven't seen them yet. What is the message and what is the purpose of the ad?"

"What the fuck have you been doing with your time?" His voice exploded. "What's this bullshit about what does the ad support? Look at the dammed ad, call Sam back, and renew the fucking ad. Get off your fat ass and get on top of your work! This isn't Disneyland, you slow-witted mule. This is the real world. This is a place of work."

His voice lowered a bit. "Are you on the fucking train or not? I think you'd better get on the train, and stay on the train, before I push you off." He hung up. My heart was pounding. I could feel my face flush. Although I'd had a strenuous workout that morning, I felt perspiration pouring out of every pore in my body. My dress shirt was completely soaked. I felt like I'd just gone ten rounds with George Foreman, and I'd been pulverized.

I went outside my office to speak to Cheryl. "Has a copy of the magazine been found yet and has a copy of the actual ad been retrieved from the files?"

"I'm sorry, Michael, I can't find a copy of the ad anywhere in the files, and I haven't received a copy of the magazine from our library. Why don't I go downstairs and look for it myself?"

I appreciated her willingness to help. "Yes, please see if you can find a copy." As the words came out of my month, I thought, "Why bother?" I had to call Sam to renew anyway, so it really didn't make any difference. Still, I wasn't going to abandon my responsibility by doing something I thought was totally unprofessional.

About half an hour later, Cheryl appeared in my office and said she hadn't been able to find the magazine. I shrugged and picked up the phone to dial Sam's number.

"Sam, its Michael Koscec calling about renewing the ad in your magazine."

"Oh good. Thanks for calling me back."

"Sam, I just want to say that I don't appreciate your calling Don when I specifically told you I was going to call you back."

Sam interjected. "Don't feel slighted, Michael. Don and I go back a long way, and there was no intention to be disrespectful. It was just a social call."

Social call, my ass, I thought. "Sam, we'll renew the ad for a year," I said. "But I'd like you to courier me your last issue. In fact, when your new issue comes out, please courier me a copy of it and put me on your subscription list. If we're advertising in your magazine, I'd at least like to get a copy. I can't find it in our library and I'd like to see it." I was too embarrassed to say we had no record or copy of the magazine anywhere in our office.

"Sure, Michael, no problem. Anything you want. I want you to be happy."

I gave Sam a purchase order number and went back to work.

Wednesday

Andy and I met in the hallway and walked to meeting room B together. Both teams were assembled and there was a wonderful positive energy in the room when we walked in. As I sat down, I noticed that David Booker wasn't there.

"Good morning. Shall we get to it? How would you like to do this, Doug?" I asked, looking at the manager of residential.

"We came to a consensus that Joe should do the presentation," he said. "Joe, the stage is yours."

Joe was affable, almost bordering on jolly. He was quick to smile, but his intelligence shone through his bright, intense eyes. "Good morning. I'm going to run through this as if Don were in the room. I'll dive right in without introductory remarks."

Joe continued, "The objectives of the residential program are to:

1. Increase sales of appliances by 10 percent, barbecues by 10 percent, and barbecue accessories by 15 percent.

2. Increase our customer base by 3 percent.

3. Increase new furnace sales and water heater sales by 5 percent.

"These numbers are based on sales increases over last year's sales. We'll achieve these targets by implementing a five-point marketing strategy:

1. A coupon book containing the incentives

2. A customer referral program

3. A contractor incentive program

4. Support advertising, and

5. Support signage."

The Coupon Book

Joe went on to explain how the coupon book would be the centerpiece of the program. It would contain one coupon for each product:

1. Twenty-five percent off any appliance, barbecue, barbecue accessory, furnace, and water heater

2. An additional 5 percent (30 percent, total) for a qualified lead

3. Name entered into a contest for a fully expense-paid holiday to Hawaii for four persons, plus $1,000 in cash for a second qualified lead. These leads would be for new customers only. One customer could not refer another customer.

4. These same prices would be offered to all new customers

Customer Referral Program

Each coupon book would contain three customer referral forms.

Contractor Incentive Program

Every contractor that brought in a new customer would receive a $500 bonus after he completed installing both a new furnace and water heater. The contractor would also receive a $150 bonus for the sale and installation of a new stove.

Support Advertising

The program was to be advertised in all newspapers for a period of two weeks.

Support Signage

A "look" would be developed for the program. Signs would be prepared for all stores, for all filling stations, and participating contractors.

Joe then went into a detailed cost benefit presentation. The margins on our products ranged between 45 and 55 percent. A couple of informal conversations with manufacturers indicated that they could give us another 10 percent discount if we achieved our sales targets. The calculations showed that the break-even for this marketing program would be achieved if we reached 50 percent of our targets. Anything after that was profit.

Andy was the first to speak. "In all my time at this company, I've never seen anything like this. It's certainly an ambitious program. How confident are you in the numbers, considering the fact that we're in a mild recession?"

"I'm confident that this program will succeed," Joe responded. "Like all marketing programs, there's always a risk. But this is a strong program in that it has so many good elements. Plus, we have the margins, despite the discounts we're proposing."

"I'd like to congratulate all of you for a job well done," I piped in. "There is one thing missing that I've been thinking about ever since we talked about this program last week. Does anyone know what I'm thinking about?"

There was silence for a moment, and then Joe spoke up, "Well, we still need a look and we need some sort of a theme."

"Exactly! I know you've have been pressed for time but have you had a chance to talk about a theme?"

Doug stepped in, "No. We've been so involved in the details, there just wasn't any time."

"My feeling is that you have the key pieces in place for our meeting with Don next week," I said. "So let's brainstorm now for a theme because we have about forty-five minutes before we get into the auto program. Is everyone okay with that?"

The next forty-five minutes were well spent. We came up with ideas like "Make yourself happy," "Be good to yourself," "Make yourself and your neighbors happy," and "Do yourself a favor." We were looking for something that was upbeat and positive, but these themes lacked something. They lacked excitement and sizzle. Then it hit us. We estimated that it would take three to four months to prepare the program. That meant we were looking at a launch date somewhere between May and the beginning of June.

"Sizzle into Summer" emerged. We all loved it. This theme beautifully covered our barbecue market, but it was also appropriate for the other products because of the "sizzling" discounts we were offering.

"I think this is great, guys!" I exclaimed. "We have ourselves a winner." I felt it in my heart, but after yesterday's discussion with Don about advertising in Sam's publication, I felt a bit gun shy. The reality was that no one could predict Don's reaction.

"I've got to admit I'm a little nervous." Andy had a concerned look on his face. "We've never done anything like this and, quite honestly, I don't know how it'll be received by Don. It's a huge program, and I'm concerned that it may be too big for us. I really don't know how Don will react. I guess the worst that can happen is that he'll yell at us and ask us to go back to the drawing board."

"I think we have to show Don what's possible, Andy." I had this sinking feeling in my stomach that the program would get scrubbed even before we had a chance to show it to Don. "We're being factual; we're showing numbers; so, I think we're speaking his language."

"I guess you're right," said Andy. "Nothing ventured nothing gained. Let's go with it."

"Okay. For Monday, can you prepare a mock-up of the coupon book? It's going to be the centerpiece around which the whole program revolves." I looked at Doug and then at Joe. "Joe, is that something you can do?"

"Yes, I think we can prepare a nice mock-up."

"Great." I looked at my watch. "It's 10:30. Let's take a fifteen-minute break before we look at the auto program." No sooner had the words come out of my mouth than George walked in.

"Perfect timing, George. We just finished the residential program review and we're taking a break before starting the auto program review. How about a coffee?" I asked, hoping I could talk to him and get to know him a little before we started.

"Good idea. I really need a coffee," he responded enthusiastically. George had a quick smile and a positive attitude. It didn't take long for me to realize that he also had a good sense of humor.

"I'm glad we finally have a chance to talk one-on-one, George," I said. We sat down in a small lounge area near the elevators. "There's so much I want to ask you, I don't know where to begin, so I'll start with *Truck and Tractor* magazine. Do you get a subscription to the magazine, and did you know that we advertise in it?"

George thought for a moment before he answered. "I'm not familiar with the magazine. I didn't know we advertised in it."

"I'm asking because I got a call from a guy called Sam Schwartz. He asked for a renewal of our advertising, and I can't find a record of a previous commitment. I told the guy I was going to call him back when I had a chance to look at our advertising program. Before I had a chance, Don calls me and tells me to renew. That's not exactly how it happened. He chewed me out first, and than he told me to renew."

George smiled grimly. "You're his new bitch," he said.

"Pardon me!" I had no idea what he meant.

"Didn't Andy tell you?" George asked.

"Tell me what?"

"Last year, Don hired and fired three marketing directors. That's fairly standard. Don fancies himself a bit of a marketing guy, and so he really likes to get involved with marketing and sales activities. Don't get me wrong. He loves to micro-manage. All of us experience a regular amount of interference, but he especially likes to get his fingers into the marketing pot."

"But why marketing?"

"My feeling is that marketing allows him to channel his creative juices more than any other function in the company. You probably know he's a chartered accountant. In many respects, the energy business is a boring business. It's meat and potatoes, and it's not glamorous. I heard that his claim to fame was his ability to grow the market share for this company."

"How did he do that?"

"When he was the manager for New Brunswick, he actually went door to door selling our products. He was driven. He was relentless. And it paid off. He was successful in growing the business down there, and he made a name for himself. This was a simple and unsophisticated business twenty years ago. Don was noticed. He was a man of action who developed a reputation for getting things done. The fact that he was a CA helped tremendously, and he rapidly rose through the ranks to become president."

"Oh God!" I jumped up. "It's almost 11:00. Everybody's probably wondering what's happened to us." We quickly walked back to meeting room B, where everyone was waiting. I had been so interested in what George was saying, I completely forgot about time.

But what he'd said was also distressing. Three marketing directors in one year! It was bad enough that Don liked to micro-manage, but why marketing? Then I remembered we had work to do. I didn't have time to feel sorry for myself.

Because I'd participated in a couple of working meetings with the team on Monday and Tuesday, I knew what to expect. I was particularly interested to see what George's reaction would be. Andy had left at the break because he had to go to another meeting. Most of the residential team remained.

Gary was chosen to present the auto program. He followed the same format as the residential team.

First he presented the objective of the transportation program: Increase conversions by 5 percent for both autos and trucks.

Then he presented a five-point marketing strategy:

1. A one-day conference and trade show

2. A customer referral program

3. A contractor incentive program

4. Support advertising

5. Support signage

Conference and Trade Show

1. Invite fleet operators:

    - Taxi companies

    - Limousine companies

    - Courier companies

    - Delivery companies

    - Freight companies

2. Invite certified conversion mechanism

3. Invite existing customers

4. Invite government officials

5. Get testimonials from existing customers

6. Invite speakers

The one-day event would be divided into two parts: the morning sessions would be the conference part of the event, and the afternoon would be the trade show with demonstrations. At lunch, we would have a "big name" speaker who was recognized and respected in the industry.

Customer Referral Program

Offer $500 to an existing (fleet) customer who provided us with a lead that resulted in a sale.

Contractor Incentive Program

Any contractor who provided us with a qualified lead would get his name entered in a contest to win a trip for four, plus $1,000 cash on a one-week Caribbean cruise on the Princess Line.

Support Advertising

The conference and trade show would be advertised in all appropriate trade magazines. Leaflets advertising the event would be sent to all current customers, all known contractors, fleet operators, and trucking companies. In addition to the advertising and the leaflets, all potential participants would get a specially designed invitation to the event.

Support Signage

A "look" would be developed for the program. Signs would be prepared for all filling stations and participating contractors.

Gary continued, "Because of the lead time that will be required to allow participants to schedule the event, and to get all of the speakers and exhibitors lined up, we're planning this event for the middle of October."

When Gary finished, he presented a detailed cost benefit analysis.

"I'm sorry Andy isn't here for this presentation, because he would have seen that the two marketing programs are quite different, and yet they share similar elements, like the customer and contractor incentives, advertising, and signage."

I turned and looked at George, "What do you think?" I asked him.

"I'm sure my guys already told you that we were thinking of doing a trade show of some kind this year, but I have to admit, in our preliminary discussions we didn't have any of this stuff that you're including. We were just thinking of a trade show, period, and," George paused for a few seconds, "and maybe a leaflet or some advertising or something, but nothing like this. Obviously this is much better. I think it's good. I'm glad to see some of my team here. Have you guys been involved in developing this with the marketing guys?"

Charles, the manager who had responsibility for transportation in George's group, spoke up, "Yeah. Harry and I have been meeting with Allan and his people to put this program together. We really like it and we think it'll produce some very good results for us. But, in addition, I think it'll also provide added profile, first, to our involvement in transportation, and second, to the benefits of our energy as a fuel. Raising the profile is badly lacking, and this will go a long way to doing it."

"Sounds good." George smiled. "When is the presentation to Don?"

"One o'clock on Monday afternoon, in meeting room A on the fifth floor," I said. "Don specifically said he only wanted three people there. My preference would be for Joe and Gary to accompany me, since they were the team leaders in their sectors. But I'll leave that decision up to you and Andy."

"Let me talk to Andy and I'll let you know."

"Can I have your answer on Friday, George? We'll need some prep time."

"Yes, I understand. I'll let you know on Friday." George thanked everyone and left the room.

I turned to the others. "I really appreciate how much you guys have done in such a short time. But something just occurred to me. I know it's almost lunchtime and you're getting tired. But something is still missing."

Joe spoke up, "What's the theme, what's the message that will tie the transportation program together?"

"Exactly. I want you to give it some thought. This is Wednesday. Let's get together briefly at four o'clock tomorrow afternoon. Bring your ideas and we'll make a decision then. I know this is transposition's problem, but if you residential guys have any creative ideas, please share them with us. You can give them to me personally between now and then, or if you can spare the time, come to the meeting. We really don't have much time to spend on this, so it'll have to be quick and dirty."

"One quick question before we break." I had to find out. "Has anybody been in touch with David Booker, or has he been in touch with anyone on the team?"

They all looked at each other. I got my answer and it wasn't the answer I wanted. We were being ignored. Merchandising was expecting us to fail and they were not going to be part of a failing enterprise.

Thursday

We met at 4:00 on Thursday to share ideas. People came prepared with slogans and themes. No one from residential showed up, although Joe and Harry had dropped off a couple of slogans earlier in the day. We went through all the suggestions and combined two slogans to come up with: "Propelling You into the Future." I liked it because it had "motion" and it tied in perfectly with the transportation sector. It only remained to be seen if Don liked it.

Friday

I started the day pedaling hard on the stationary bike to work off the anxiety that came over me when I thought about the regular Friday meeting. Although I had been to only one senior management meeting so far, I dreaded the thought of having to go to the boardroom again. It wasn't only anxiety that consumed me, but also fear. The perspiration on my forehead began to flow freely down my nose and along the sides of my face, collecting at the end of my chin. This was the most strenuous workout I'd had since I'd started the previous Wednesday. When I finished, I was exhausted, but I didn't feel any less anxious. I showered and dressed and went to my office. I was frustrated by the fact that I didn't know how to prepare myself for these meetings. There was an agenda, but as I'd seen at the first meeting, the agenda didn't matter much. We never knew when that sidewinder missile would hit. We knew where it would come from. We just didn't know when it was going to be fired and where it was going to land.

I sat down in the same chair as I had last week. I decided there was no point in moving further down the table. When the missile was fired, it would hit its target regardless of where I sat. Andy walked in and sat beside me.

"How are you this morning, Michael?"

"I'm okay," I said.

"I saw you working out again earlier. Is this your regular routine now?" The fitness center was located in the basement by the back entrance to the building. A floor-to-ceiling glass wall separated the fitness center from the foyer. Anyone using the back entrance could see who was using the center. And anyone coming to work between 6:00 and 7:00 in the morning saw me working out. Don usually arrived between 6:30 and 7:00, and most of the vice presidents and directors arrived between 7:00 and 7:30.

"Yes," I told Andy. "I like working out in the morning before work. When I discovered you had these great facilities, I decided to take advantage of them. I'm surprised more people don't use the fitness center."

George waved to us as he walked into the room and sat on the other side of the table, a few chairs to our left. The boardroom was filled by 8:55 when Don walked in with Teresa in tow.

"Good morning, everyone," he said. "Let's get started with the agenda. Carry-over items from last week. Jack?"

"Yes, Don. The inspection schedule has been integrated into the technicians' work schedule."

"I knew those technicians were underemployed, those lazy little weasels," said Don. "At least now we'll get our money's worth." He held up his agenda. "Next item: transportation. George, I was looking at the transportation sales numbers and noticed that they haven't budged from last year. What the hell are you guys doing? No, let me rephrase that. I could fire the whole lot of you and no one would notice!" Don's voice started to rise, but he didn't get up from his chair or slam the table.

"Actually, Don, I haven't had a chance to meet with you to let you know that Michael has included transportation in his new marketing program, and we'll be presenting this to you on Monday. The program is designed to really boost transportation sales."

"Well, let's put the agenda to the side for a moment." Don turned to me. "Michael, the floor is all yours. Enlighten us with your transportation marketing program."

"Don," I said, "I didn't bring any of the materials with me. We weren't scheduled to meet with you on the new marketing program until Monday afternoon, but the—" Don got to his feet and cut me off before I could finish my sentence. I had been going to say that the centerpiece of the transportation program was a one-day conference and trade show.

"What the fuck does that mean?" he shouted. "I don't give a shit if we were scheduled to meet on Monday or not! If I ask you for something, I expect you to give it to me. I don't give a flying crap if you have the materials or not." He made a sarcastic face, rolled his eyes, and then glared at me. He sat down in disgust, exhaled loudly, and continued: "Michael, you're not getting off to a good start. If this keeps up, your days here are numbered. Marketing guys like you are a dime a dozen. Unless you pick up your socks, you're going to be history. All right. Back to the agenda. What's next? Harry ..."

Everything was a blur. Don's words trailed off, and I didn't hear a sound for what seemed like an eternity. I really blew it, I thought. I'm doomed. Why didn't I immediately launch into the program description instead of that stupid excuse? What the hell was wrong with me? I knew better than that. I'd set myself up and opened the door for him to come right in and knock me out. I'm such an idiot! I slowly began to tune back in to the meeting.

"Harry, you dim-witted idiot! You should know that by now! How long have you been here? Can anybody else tell me why we schedule our deliveries in this staggered fashion? Anybody?"

Stan Ward started to speak. "By staggering the deliveries, we're placing ourselves in a better position when winter comes and when our customers start to heat their homes. This way, people run out of energy at different times. This allows us to maintain an even level of distribution and smooth out potential peaks and valleys. We can only do this by purposefully not filling up some tanks during the summer when they're almost empty."

"Are you listening, Harry? Can you remember that, Harry?" The emphasis landed heavily on the last Harry. "I'm surrounded by imbeciles," Don said in a voice of resignation. "Let's move forward."

The rest of the meeting continued in the same vein. It was pure agony. I thought a meeting couldn't get any worse than last week's. Boy, was I wrong. This meeting surpassed the previous level of nastiness by a wide margin. The weekend couldn't have come fast enough for me.

Week Three

Monday

I started Monday morning by following my usual routine. I arrived at the fitness center at 6:00 and I was feeling relaxed. I'd had a wonderful weekend. My wife and I had celebrated my birthday with friends and we'd had a fantastic time.

During my workout, I focused on the marketing program presentation. Don only wanted three people at the meeting. Both Andy and George wanted to be there, which meant I couldn't bring Joe and Gary, the two team leaders. But after only two weeks of exposure to Don, I wasn't prepared to argue with him on this point.

At eight o'clock, I met with the residential team. I'd spent most of Friday after the senior management meeting preparing both the residential and transportation presentations for Don. I wanted to do a dry run of each program on Monday morning to make sure there were no loose ends. The residential team showed me the mock-up of the coupon book. They'd done a great job. They'd not only put the appropriate information on each coupon but they'd also created a design for the cover, carried the design on each coupon, and added a bit of color, giving the coupon book a look of authenticity. It was comforting the see the actual props that were going to support the presentation. The blueprint for "Sizzle into Summer" was ready.

At 10:00, I met with the transportation group. To my surprise, they had also prepared a mock-up design to convey the "Propelling You into the Future" theme. They'd designed a logo that incorporated a propeller to give the feeling of propulsion. I was impressed. An hour later I walked out of the meeting, and the blueprint for "Propelling You into the Future" was completed.

One-thirty seemed to take forever to arrive. Admittedly, I was nervous. I met Andy at 1:20 and we walked up to meeting room A. George's office was on the fifth floor, and he was already in the room when we arrived. Don was late. He arrived at 1:40, making the wait that much longer. My stomach was in knots.

"Good afternoon, men," Don said in a matter-of-fact way. He was wearing his customary navy blue pinstriped suit. "Let's begin."

To my surprise, George started to speak. This was not the way the meeting was supposed to play out.

"Don, I'd like to start and tell you about the transportation program."

Don immediately interrupted. "No. I want to hear about the marketing program from Michael."

I started right in without missing a beat. "We've created two programs: one for residential and one for transportation. The objectives of the residential program are …" I continued to describe the residential program. I didn't know how often I was going to be interrupted with questions. I spoke for about five minutes, and shortly after I introduced the coupon book, I stopped and asked Don if he had any questions.

"If I have any questions, I'll ask. Just keep moving forward."

My sales experience had taught me that you always stop to ask for questions and to solicit a "yes" as you run through your sales presentation. If you get three "yeses," you're pretty well assured of a sale. I saw this presentation to Don as a sales presentation. I needed to get him to accept this marketing program. We'd invested a great deal of time and effort in creating it. And besides, my reputation was on the line. No, not my reputation. My life was on the line. The stakes were the highest I'd ever faced in my whole working life. I knew this was do-or-die. And Andy and George knew it, too.

Unfortunately, Don was not going to allow me to use my own playbook. I had to play by his rules. So, I continued right through the residential presentation.

"This concludes the 'Sizzle into Summer' residential program," I said when I'd finished. I didn't know whether or not to go straight into the description of the transportation program. On the one hand, it didn't seem right to continue. On the other hand, I wasn't going to ask Don if he had any questions. I had learned my lesson. So I decided to frame the ending in a slightly different way.

"Do you want to discuss the residential program before I begin the presentation on the transportation program?"

"No. The program is clear. I like it. I like all the elements of the program and I like the theme. I want you to change the launch date from June 1st to May 1st. Andy, what do you think?"

"I like it as well, Don. It's really comprehensive."

I was disappointed that Andy didn't object to the change in the launch date. The new date was going to stretch us to the limit. As much as I wanted to say something, I decided not to push my luck.

"I have one question," Don said. "Has Craig or any of his people been involved in the program? Has David seen it and commented on it?"

Oh boy, I thought. Was this a trap? I didn't understand the politics, and I wasn't sure how to answer. Luckily, while I was thinking about it, Andy spoke up. "Craig was informed about the program, and he asked David to provide feedback to the team. Michael spoke with David and provided him with the information about the team meetings and the feedback sessions. David was not able to attend any of the meetings, and so as far as I know, merchandising is not aware of the program details yet."

"What the fu—… oh, never mind." Don stopped himself. "Good, Michael. Go ahead with the transportation program."

I felt like someone had just lifted a thousand pounds from my shoulders. I quickly glanced over at Andy. He had a slight smile, and I could see the tension melting from his face. I launched into the transportation program, not veering from my script.

When I finished, I looked at Don. This time, I decided not to ask any questions.

"I like the transportation program. It's better than the trade show you were planning, George." He was looking at George, expecting a response.

"Yes, I agree, Don. This is a very good program—much better than a straight trade show."

"Okay. There is one thing I want you to change, Michael. I want you to move the launch date of the transportation conference and trade show from the middle of October to the middle of June."

I looked over at George and he looked at me. There was silence. The middle of June didn't provide enough lead-time to organize a proper conference and trade show. Obviously, George wasn't going to say anything.

I decided someone had to say something, "Don, the middle of June gives us a very short time to line up the caliber of speakers we want and to line up all the exhibitors. That's why we were planning the event for October."

"I want this marketing event in June," Don said in an even voice. "And if you aren't prepared to do it, Michael, I'll find someone else who will."

There wasn't much else to be said. I knew why George hadn't spoken up.

"I like what I see," said Don. "Go ahead and put the programs together, and keep me informed of your progress." He stood up to leave the room. But there was one question I needed to have answered.

"Don, I'd like to ask you one quick question before you leave. I understand that Campbell Marshall is the advertising agency of record for our company. Do you want us to use them, or is there another company you'd prefer? We need to engage the advertising firm as soon as possible."

"Use Campbell Marshall," he said. He turned and walked out of the room.

Somehow, Don had managed to take away my brief feeling of triumph and replace it with concern and frustration. I had just finished presenting two very good marketing programs that Don had liked and approved. Instead of feeling happy, I felt depressed. We were going to have to work around the clock to get the two marketing programs prepared by the time Don wanted them. We had to accelerate the residential program by one month and the transportation event by four months! It would be almost impossible to arrange a successful transportation event for the middle of June. I felt like I was being set up for failure. I also didn't know what the fallout from merchandising was going to be.

"I guess this is a good news-bad news story." I looked at both Andy and George. "What do you think about the lack of participation by merchandising?"

Andy sighed. "I see this as a major win for us. There was total acceptance. Don didn't ask any sticky questions. He didn't criticize. He actually said that he liked it. Personally, I'm elated. As for merchandising, my read is that Craig and David thought that Don had set Michael up for failure. They thought that Don was not going to like whatever we came up with because of the hard time Don had given Michael at each of the senior management meetings. Merchandising didn't want to be associated with a flop. They didn't count on a win. So, I think we're okay, but I think merchandising is screwed. We'll just have to wait and see how Don will go after Craig and David."

"I totally agree," George added. "It was a gross miscalculation on their part. As for the accelerated timetable for the conference and tradeshow, we'll just have to pull out all the stops to make it happen. The important thing is that Don liked it and he wants us to make it happen."

"I'm really going to have to depend on your guys, George. I'll need 100 percent cooperation. My guys are not going to be able to do all the work to get the event launched on time." But I was still concerned about meeting Don's deadlines.

"Let's get together with both teams in half an hour," said George. "We need to get started."

As we walked through the corridors, I could see individual heads peeking around their office screens, anxiously waiting to hear the news. I gave them all thumbs-up. I went into the coffee room and poured myself a coffee. On the way back to my office, I stopped and briefly talked to everyone.

When we gathered at 3:30, Andy spoke first. "We had a great meeting with Don. He liked both programs and he gave us the go ahead. There was no acrimony. No negative comments. He liked what he heard. So, it's full steam ahead for both the residential and the transportation programs. However, it needs to be fuller steam ahead: the residential launch date will be May 1st and the transportation date will be mid-June. That's why we're meeting now." Andy looked at George.

George continued. "I've given Andy my commitment that this will be the top priority for you guys, providing Michael's transportation team all the support they need in order to meet the mid-June launch."

Gary put his hand up. "George, I appreciate that this is a tight schedule, but we have a lot of our own work to do that's also a priority. We're being placed in an impossible position."

"What else is new?" George said.

"Gentlemen." I felt that I needed to step in at this point. "I know we're being placed in an impossible situation. But we've been dealt our cards, and now we have to play the hand we've got to the best of our abilities. What I'm going to say applies to both the transportation and residential teams. I want you to construct a critical path chart that lists everything that needs to be done. Beside each activity, show the length of time it will take to complete. Show actual start and end dates. This has to be real. Then identify those activities that can be done in parallel—this is especially critical for the transportation program. Be creative. Overlap as many activities as possible. This is the only way we'll be able to compress our original timeline by four months and reach our May 1st target for residential and mid-June target for transportation.

"One other thing," I said. "Send merchandising every team meeting notice. Make it clear that they're invited to every meeting. Make sure that David Booker is invited, and copy Craig at all times. Keep them in the loop every step of the way. Send them copies of all minutes of meetings. Keep a complete paper trail. If they can get involved directly and help, that will be great. If they can't, that's okay, too. But I think at this stage, we need to move forward on the assumption that we'll have to carry the load." I hated having to cover my behind, but it was clear to me that we had to do it.

"I'd like you to work on this tomorrow," I continued, "and have it completed by the end of the day. We'll meet Wednesday morning to review the transportation critical path schedule, and meet in the afternoon on the residential schedule. In the meantime, I'll call Campbell Marshall and arrange a meeting to go over the program. I'll also talk with the regional directors to give them a heads-up. Any questions?"

A few good questions followed that brought greater clarity to the work that had to be completed.

Tuesday to Thursday

The next few days flowed seamlessly, one into the other. The flow was only disrupted by the Friday morning senior management meeting.

Friday

In my mind, I now referred to the senior management meetings as the "dreaded" Friday morning meetings. My first two meetings were indelibly etched in my mind. I still had the scars. I didn't know how long it would take for them to heal, if ever. This morning, I sat down in my usual spot and braced myself. I had recently seen the film *The Longest Day*, the story of the D-Day invasion of Normandy. It was at least fifteen years since I last saw it. I had completely forgotten that the movie showed scenes from the Allies' perspective and from the German perspective.

I don't know how much of it was historically accurate, but I was especially fascinated by the dynamics among the top German generals and Hitler. Just as the invasion began, one of the generals, played by Curt Jurgens, called Hitler's office to get his approval to release a Panzer unit to the beaches of Normandy. Hitler had complete control over the Panzer divisions, and was the only one with the authority to order their movements. Apparently, Hitler had taken a sleeping pill before the invasion started, and his aid refused to wake him up. The requesting general was incensed.

The invasion was well underway when Hitler woke up. He was in a foul mood and immediately launched into his agenda, firing off orders and having temper tantrums. The attending generals and aides were too afraid to ask Hitler about sending his Panzer unit to the Normandy beaches. They were even more afraid of Hitler than they were of the enemy. As I watched these scenes, I wondered how much of what went on in the company was because Don ruled with fear and an uncontrollable temper. How much was being held back from Don that could help the company increase revenues and profits, because employees and managers feared him?

Don began the meeting by talking about the two-day regional managers meeting in February. He was developing the agenda for it. He also dropped a bit of an organizational bombshell. I discovered later that Don liked tinkering with the organizational structure, and he announced new organizational changes about every four months. "I'm establishing a new 'high performance team' that will be responsible for growing our business," he said. "It'll be made up of three people who will report directly to me. They are Ian Foster, Bill Collier, and Barry Wilson."

Ian Foster was the regional director of Western region. Bill and Barry were up-and-coming branch managers: Bill was from Keswick; Barry was from Chicago.

"I'll reveal greater details about this high performance team at the regional managers summit. I'll also have Michael do a presentation on the new marketing program. Although the agenda is still soft at this stage, I see one hour being given to you, Michael, on the morning of the first day."

This was totally unexpected. I looked over at Andy, and I could tell from the expression on his face that this was news to him as well. Nevertheless, I was thrilled. This was confirmation of the fact that Don really liked the new marketing program. It was a real vote of confidence. This was the best I'd felt since starting at Northern Energy. I saw my star slowly rising. Or was it?

After Don finished talking about the regional managers summit, he turned to Marcia and Craig, "Let's look at the new HR policy for the store employees, and the store employee sales training program. I don't care which one you begin with. Who would like to start: Craig? Marcia?"

Marcia began, followed by Craig. Don listened silently for a while. Then, suddenly, he shot back with a barrage of expletives, most of which I'd never heard before. He spewed bullets from his mouth in rapid succession like a Gatling gun. They pierced the body with multiple volleys and they mowed down his victims. They were immobilized and they were silenced.

I looked at Don when he was finished. The expression on his faced shocked me. It was not an expression of anger, frustration, or hatred. He was spent, but happy. His face was flushed, but content. I realized this kind of tirade fed Don. He had the same expression on his face as an athlete who has just won the gold medal at the Olympics—tired, but overwhelmed with joy. The athlete pushed the limits of possibility and clutched victory in his hands. The endorphins were pumping, and he reached a high of pure ecstasy.

I walked out of the boardroom at the end of the meeting in a state of shock. I thought these Friday morning meetings couldn't get any worse, but this meeting had reached a new low. I now saw that I was working for a madman. How do people like Don become company presidents? Was this Don's way of venting his anger at Craig for not participating in the development of the residential marketing program? Would this be the end of it, or would Don continue to torment Craig and David?

Week Four

When I'd asked my staff to prepare a critical path for the programs the previous week, it occurred to me that I'd better do one for myself as well. I was juggling too many balls at the same time, and I knew I couldn't afford to drop one. All the work had restrictive deadlines. I spent Monday morning preparing my own work plan and critical path to the middle of June, the launch of the transportation event. When I completed it, I was overwhelmed. I sat back in my chair with my hands clasped behind my head. I stared out the window for a moment. It was a gray January day. Even though it was almost noon, there wasn't much light.

As my eyes scanned from the window to the painting on the wall, I noticed that the corner join running from the ceiling to the floor should have been straight, but it wasn't. It looked jagged. I looked at the opposite corner. It wasn't straight either. The line looked jagged. I thought this was strange, but then my telephone rang and I forgot about it. It was Bob Marshall from Campbell Marshall, our advertising agency, returning my call. I briefly spoke with Bob and explained what was happening. He agreed to come to the office Tuesday morning to hear about our programs in more detail.

Cheryl arranged a conference call with all the regional directors for Monday afternoon at two o'clock. This was going to be my first opportunity to talk to them about the new marketing programs. Shortly before two, Cheryl walked into my office with a news bulletin. Don had had a 1:00 o'clock meeting with the new public relations manager, the replacement for the one he'd fired at the beginning of January. Apparently, after half an hour with Don, she just stood up and quit. She walked out! No one knew the details.

There was some speculation that Don had played his usual "threat of firing" card, and she'd quit after hearing it the first time. Everyone, including me, shared the same thought: "Bravo. Good on her for having the balls to stand up. She wasn't going to be bullied."

I asked Cheryl, "Does a week go by at Northern Energy where Don doesn't do something to create waves, or controversy, or pain? Does he ever take a holiday from stirring the pot?"

I knew the answer before Cheryl's emphatic response. "No! Now, you'd better get ready for your teleconference."

The regional directors responded enthusiastically to the new marketing programs. I promised to provide them all of the program pieces well in advance of the official program launches. During the conversation, it occurred to me that we would have to provide the regions with a program instruction book and some training. In the rush to prepare the program outline, we hadn't discussed this and that it would have to be added to our critical path schedule.

The meeting with Campbell Marshall on Tuesday went well. Over the years, they had not had an opportunity to be involved in a fully integrated marketing campaign with Northern Energy. Their work was restricted to preparing the copy and artwork for the various pieces in the marketing playbook. They were looking forward to participating in this program. Bob mentioned that he had been after Don for years to do something more than the local print ads for the various products. But Don hadn't been interested. He liked the playbook and he was going to stick with it.

Bob liked the residential program. He suggested that the program be supported by two or three radio ads that would be played in the recreational areas across the country. He said that radio advertising was relatively inexpensive at the smaller rural radio stations. This was where a large part of our program was targeted. The idea sounded good. Bob promised to come back with a cost estimate in a few days.

The week was packed with work. On Wednesday, we were advised that Don was going to be in Chicago for the rest of the week and wasn't returning until early evening on Friday. There was to be no Friday morning meeting. You could almost hear a collective sigh of relief across head office. The whole atmosphere lightened. Nevertheless, I still felt a great deal of pressure. I had to start work on the Maritime marketing plan. I had less than four weeks to complete it.

Week Five (The beginning of February)

Although time frames were tight for both the residential and transportation programs, the two teams developed work plans that were going to complete the programs on time. Campbell Marshall allocated dedicated resources to us. Everything was in place. Even staff from merchandising was now involved. David, their director, personally attended all the team meetings. He kept Craig apprised every step of the way. I was informed that the merchandising folks were active and helpful participants. This was good news because we needed all the help we could get. I felt that a gigantic machine had been created, and it was now well oiled and running at peak performance.

After a month at Northern Energy, I had dropped fifteen pounds. My early morning workouts were paying off. I also noticed that straight lines continued not to appear straight. They seemed to be getting more jagged with each passing day. I had never had problems with my eyes in the past, but now I was worried. I didn't know any ophthalmologists, and it had been at least two years since my last physical, so I decided to kill two birds with one stone and call my family doctor. Mike Madison had been my doctor for the past fifteen years. I called him and arranged a physical for Thursday at lunchtime.

Tuesday

Tuesday morning started out like any other morning. Traffic to work was a little heavier than usual, even though it was only 6:00 in the morning. It had snowed during the night. There was over two inches of fresh snow on the ground, and many of the streets had not been plowed, but I was still able to get to the gym in reasonable time and have a good workout before settling at my desk at 7:30. I wasn't there for more than five minutes when my phone rang.

I knew who it was from the call display. "Good morning, Don. How are you?"

"I'm well, Michael." After a short pause, he began to speak. "When I was coming back from Chicago last Friday, I decided that we needed a new company logo. I'd like you to approach three different companies to develop several concept designs that we can look at. One of the individuals I'd like you to contact is Peter Brown. Mary will give you his number. You may want to ask Campbell Marshall as well. And dig up a third company somewhere."

I was completely caught off guard. We were starting to prepare the artwork for our marketing campaigns. Many of the pieces included the company logo. We couldn't stop the work on account of the logo. But if Don wanted a new one, our current logo would be obsolete when the programs hit the street. So much for our well-oiled machine. I didn't know what to say.

"Okay," I began slowly. "And when would you like this process to begin?"

"Right away. Otherwise I wouldn't have asked you now."

"But you know we've started to prepare the artwork for the new marketing campaign. Some of the pieces include the company logo. How are we going to deal with this if we end up with a new logo at program launch?"

"You'll have to cross that bridge when you get to it. I think the company needs a more modern look. We're too staid. I'd have thought you'd be happy, in light of this great new program you're creating. We're moving forward into a new era, and we need a new look to go along with the great new things we're going to be doing. I mentioned this to the three new guys who'll be part of the high performance team I'm creating, and they're all for it. If we're going to be a new company in a new era, we'll have to project the right image."

"If we're going to ask these companies to prepare concept logos," I said, "they'll be providing us with a product even though it'll be in concept form. Can we offer them remuneration for their work? I'm thinking especially of the two unsuccessful entrants."

"I think that's fair. What do you think is a reasonable amount?"

"What about $1,000 each?"

"Let's do it."

I didn't believe my own ears as I heard myself speak. "Okay, Don. I'll call Peter and Bob, and I know another talented graphic artist. I'll ask them to prepare concept logos."

"Thanks, Michael." Don hung up. The peace of the previous week vanished in one short telephone call—another project to fit into my already crammed work plan.

I immediately called Mary to get Peter's coordinates. I soon learned that Peter was Don's nephew. I couldn't pass on any more work to my staff, so I called Peter, Bob, and Fred Thompson myself. Fred was a neighbor and a very talented graphic artist. He'd worked for IBM for twenty years. Four years earlier, Fred had decided to go out on his own. He was extremely talented, and after four years, he'd managed to build a solid client base. He never looked back.

I called the three individuals and asked them if they'd be interested in submitting draft concept logos by the middle of the month. I told them about Don's image of a more progressive look for the company. I also mentioned that the two runners-up would be compensated $1,000 for their efforts. We agreed that they would have draft concept designs in fourteen days. I asked Cheryl to book three half-hour meetings and to coordinate this with Mary. Don had to be present for these presentations. I also asked Cheryl to find out who else Don wanted at these presentations.

Fred agreed because we'd been neighbors for fifteen years and had developed a trusting friendship. Peter agreed because of his personal relationship to Don. I called Bob last.

"Hi, Bob," I said. "You're not going to believe why I'm calling, but here goes. I got a call from Don and he wants a new company logo. He wants a more modern look and a new company image to reflect the more innovative things we're doing with our new marketing programs and with his new high performance team. He told me to start on the process now. He wants me to contact three companies to prepare concept logos. You're one of the companies he suggested."

Bob laughed. "It's never dull at Northern Energy, is it?"

"My concern," I continued, "is that you've already started work on the two programs. We don't have the luxury of time to slow things up for a new logo. Even if concepts are developed in fourteen days, there's a lot more time needed to move from concept to final design."

"You're right. So, let's deal with one thing at a time. When do you want the concepts to be completed?"

"The other two companies agreed to have something in fourteen days."

"Can you tell me who they are?"

"Sure. One is my neighbor, Fred Thompson, and the other is a guy by the name of Peter Brown. He's up near Lake Simcoe."

"I don't know Fred, but I do know Peter. He's Don's nephew. In any case, we can prepare some concepts in fourteen days. With respect to the artwork that we're working on now, I suggest that we leave out the logo on all our material. We'll use the company colors as planned. Regardless of the logo, it'll take a long time to change the company colors. There are several hundred trucks, over 125 company stores, not to mention all the filling stations. A color change of that magnitude will have to be budgeted and would probably take a couple of years. And there's a high probability that it won't happen anyway. So I think we're safe staying the course. And remember, we came up with a concept look for the 'Sizzle into Summer' program, and this can still continue, with or without a logo. And don't forget, Don approved the look when he approved the program. So we'll continue with the artwork without the logo."

Bob had made my day. "Thanks, Bob. What you said makes a lot of sense. I appreciate your wisdom. You managed to reduce Don's last request from a major problem to a small, manageable glitch. We'll talk soon."

I called Don and left a message, letting him know that three firms had been lined up and would have concept logos prepared in fourteen days.

I could breathe a lot easier now and returned to my work—the Maritime marketing plan. Soon, I'd have to tackle my presentation for the regional managers summit.

Thursday

I arrived at my doctor's office a few minutes before noon. Luckily, there weren't too many people in his waiting room. The nurse called my name after about ten minutes.

"Michael," my doctor said when he saw me, "you haven't been here for almost two and a half years."

"I know," I said. "It's been a while since I had a physical. I've noticed lately that my eyes are getting bad, and I'll probably need a referral to an eye doctor."

"Let's check your weight, and then we'll see what your blood pressure is doing." He motioned me over to the scales. "A hundred and eighty pounds," he said. Close to your running weight from a few years ago. That's great."

He connected me to the blood pressure cup and started pumping. "Blood pressure's good, too," he said. "A hundred and twenty over eighty."

He checked my lungs, abdomen, ears, throat, eyes, reflexes, and then asked me to get dressed.

"Everything looks good, Michael. You're in great shape. I do want you to get blood work and an EKG, though. Now tell me about your eyes."

"When I look at a straight line, it doesn't appear straight. It appears jagged. And it seems to be getting worse. I also noticed that things are starting to look a little fuzzy when I read."

"You may well need to see an ophthalmologist. However, I'm concerned about the jagged lines. Tell me what's happening in your life? How's your family? Your job?"

"My family is great. My wife has gone back to university and is loving it. Michelle and Samantha are doing well in school. In January, I started a new job. It's a great position, but there's a lot of stress. The company president is a maniac."

Dr. Madison interrupted. "What do you mean, he's a maniac?"

"First of all, he's a bully and he has a quick temper. He picks on everyone equally, and not just me. Secondly, he micro-manages. He comes up with new ideas that interfere with ongoing work. And thirdly, he wants things done right away. He demands unrealistic deadlines."

"Can you give me examples, to help me understand better?"

"At our Friday morning senior management meetings, you never know who he'll berate and yell at. It could be me one week and someone else the next. He shouts and swears, and will even bang his fist on the boardroom table." I could feel myself starting to get emotional as I spoke about my work situation. "If you question something he asks you to do, he threatens you with firing. These are not idle threats, either. My first day at work, he fired three people. Apparently, he likes to start the New Year this way to send a signal to everyone that he's not to be trifled with."

The floodgates were opened and I couldn't stop. Dr. Madison just listened. "We're working on a new marketing program that he asked for and that he likes. But he's placed unreasonable timelines on the program. There was no discussion. We must comply with his wishes or we're out the door. So all of my team, me included, are working eleven, twelve, and more hours a day, just to keep our heads above water. I can handle hard work, but I find it difficult working in an environment where there's so much uncertainty. I'm working on shifting sand, and I never know what else will be thrown at me."

"Michael, get these tests done at the lab across the street as soon as you can, and then, let's get together when I have the results. How soon can you go to the lab?"

"I can't go tomorrow, but I can do it Monday morning."

"If you go Monday morning, I'll have the results by the end of the week. Make an appointment to see me as soon as possible the following week. My first impression is that your eyesight is deteriorating due to stress, but let me look at the test results first, and then I'll be in a better position to talk to you. See you in about ten days."

When I left the examination room, I made an appointment for Monday, eleven days later. My appointment was the day before the two-day regional managers summit.

Friday

Another dreaded Friday morning had arrived. To my surprise, the meeting was rather civilized. There were no outbursts, no upsets. Don seemed in a rather good mood. We went through the agenda in an orderly fashion, and the meeting ended on a high note.

"On the Tuesday night of the regional managers summit," Don told me, "instead of our typical dinner in a private dinning room of a restaurant, we were able to secure a box for a hockey game. We'll have food and drinks provided in the box."

That was the end of the meeting. I couldn't believe it.

Week Six

I only had two and a half weeks left to complete my Maritime marketing plan. I was able to do a little bit here and there, but I needed a dedicated block of time, where I could immerse myself in all the data and information I had gathered to conduct a proper synthesis, and to develop a strategic marketing direction. I found that I couldn't be actively involved in a lot of activities, and then pick up the marketing plan where I last stopped and start writing again. Although I had a number of meetings arranged with Campbell Marshall and my marketing teams this week, I managed to block off a complete half-day for the marketing plan on each of the next five days. Because of the pressures at work, I'd spent half a day working on the plan on each of the past two Saturdays. My deadline was approaching too quickly, and I felt myself panicking.

Just before burying myself in the plan, I quickly riffled through the time-sensitive in-basket that Cheryl had set up for me. I came across the invoice from *Truck and Tractor* magazine. I'd completely forgotten about Sam Schwartz and *Truck and Tractor*. I walked over to Cheryl's desk.

"Did we ever receive a back issue of *Truck and Tractor* magazine?" I asked her.

"No, I didn't see it come in yet."

"Sam Schwartz promised he was going to courier me a copy; that was—what?—four weeks ago? Call Sam and tell him I'm not going to pay his invoice until I see the December issue of the magazine. Tell him to courier it to me right away and to include the January issue." I didn't have a good feeling about this, but I didn't have time to do any further investigation. I decided I'd take it one step at a time and see how it evolved.

I left the office at 7:30 Wednesday evening. This had now become the norm. As I left, I was feeling a bit more positive. The marketing plan was starting to come together, and we were on track with the two marketing programs. Just before Cheryl had left for the day, she told me that Don was going out of town tomorrow morning and was not coming back until late Friday night. This meant that the Friday morning meeting was cancelled. Mary was going to distribute the final agenda for the regional managers summit. The meetings were taking place at the Embassy Suites Hotel on Highway 7.

Once again, I felt a sense of relief on hearing that the Friday morning meeting was cancelled, despite the fact that the last senior management meeting had been relatively benign. It gave me more time to spend on the marketing plan.

Week Seven

Monday

By the end of the previous week, I had managed to break the back of the marketing plan. I was in a position where I could send key pieces of the plan to Tom McAllister, the regional director. It was important to get his input on the strategic direction. I had to get his buy-in before completing it and giving it to Don.

Shortly before noon, I went to my doctor's appointment.

"Hello Michael," said Dr. Madison. "I've got your test results. First, let me take your weight and blood pressure again."

I stepped on the scales. "Looks like you've dropped another two pounds. You're down to 178." He strapped on the blood pressure cup and pumped. "Still normal," he said.

Dr. Madison sat down at his desk and indicated I should sit in the chair beside him. He opened up my file folder. "I've gone over all of your results: cholesterol is normal; triglycerides are normal; and all the other blood tests are normal. Your EKG is fine. Quite frankly, Michael, you're in great shape. Clearly, your exercising is paying off. I called a colleague of mine, Dr. Kastner, to discuss a theory I have about your eyesight. He agreed with me that there's no need for you to see an ophthalmologist. You're in perfect physical health, but you're currently under a great deal of stress at work."

I nodded in agreement and he continued. "Stress is an interesting thing, Michael. A periodic high level of stress is not harmful; in fact, it can be healthy. Short-term stress can provide positive energy and motivation. However, prolonged high levels of stress with no end in sight can be harmful. The stress can be caused by continued high volumes of work and tight timelines. Typically, this type of stress is externally imposed. You feel helpless because you have little or no control over the work conditions that are being thrust upon you. It's sink or swim. Add to this what you described as a maniacal boss who has a bad temper and is a bully, and you have a lethal combination."

I interrupted. "You described my work situation perfectly."

Dr. Madison continued. "A prolonged high level of stress will do several things. It will weaken your overall immune system. We didn't talk about this ten days ago, but stress will affect your sleep patterns and your sex drive. It will also act like a heat-seeking missile. The stress will seek the weakest part of your human system and attempt to destroy it first. In a large number of people, this happens to be the heart, so people get heart attacks. The heart is vulnerable because people are out of shape, they don't eat well, etc. In your case, you're very healthy. You're also physically fit—your level of fitness is in the top ten percentile. Plus, you're emotionally and spiritually fit. You have a solid marriage and a happy family life. I know you have a strong faith. The weakest link in your system is your eyes. Your eyesight is deteriorating due to stress. There's nothing an ophthalmologist can do for you. Relieving your stress is your only hope of restoring your eyesight."

"What you say makes sense, but how can I relieve my stress?

"I don't mean to sound pessimistic, but that's going to be difficult. There are stress management workshops that will teach you how to deal with stress. However, some of the stress management techniques involve physical exercise as a way of relieving stress. You're already doing this. Others involve meditation as a way of gaining peace. Again, I know you spend time in prayer and meditation each morning. Basically, you're doing the two most important things to relieve stress, but you're slowly succumbing to it. I could prescribe a medication, but I'm afraid that, in your situation, it may do more harm than good because of the possible side effects. For example, you obviously have to be sharp at work. The meds will take away that edge. I think the side effects will get in your way and reduce your effectiveness. My recommendation is for you to find another job if that's at all possible."

I sat in silence, internalizing everything Dr. Madison had told me. I thought of Art Bertram, who had died on the job at Hydro. I replayed in my mind the way he'd collapsed in front of me from a heart attack. I saw Al Davis bending over him, giving him mouth-to-mouth resuscitation. I saw the wet trousers that signaled his death. Those images will never leave me. Art had been under a great deal of stress for a long time. His heart had been his weak point. But I had never made the connection between my stressful job and my deteriorating eyesight.

"Thanks, Dr. Madison," I said. "You've given me a lot to think about. At least now I can see how my job is affecting me personally."

"Michael, I'd like to see you again in a month's time. Will that be possible?

"Yes I'll make it possible. Thanks, doctor. I'll see you in one month."

I left for the office; I needed to prepare for tomorrow. When I got there, Cheryl told me that the three companies I had asked to prepare concept logos had called to say they'd have something to show us as scheduled. I asked her to confirm the three thirty-minute sessions with Don for each logo presentation.

Tuesday

We assembled at the Embassy Suites at 8:30. There were about forty people in the room. The regional managers summit included all vice presidents, directors, regional directors, branch managers from the larger branches, and selected head office managers who had an active role to play. Doug, my manager of residential marketing, accompanied me. Allan, the manager for transportation marketing, was also there.

I was scheduled to speak at 11:00. I had an overhead presentation prepared that included photographs of the preliminary artwork of all the marketing materials that were going to be used. Not even Don had seen this material. I'd waited until the last minute to put this portion of the presentation together, so that I could get the latest available artwork from Campbell Marshall. It was still at the conceptual stage, but it gave a fairly good sense of the look of the residential program. I also had a few pieces for the transportation trade show. But the residential was the program with the higher profile, and, as such, there was considerably more material to support it in my presentation.

Don opened the meeting with the customary welcome remarks. He went over the agenda. The summit would end on Wednesday at noon. There were quite a number of technically oriented items on the agenda. Four agenda items were non-technical: Don's presentation on the future of the company; his presentation on the high performance team; mine on the marketing program; and Ron Johnston's, the director of finance, on the financial health of the company. I noticed that a cocktail party was planned at the hotel between 5:00 and 6:30. There was no mention of dinner or the hockey game. Mary had sent me a ticket for the box a few days ago, and I hadn't given the game a second thought.

When Don concluded his welcoming remarks, he spoke about the future of the company, his expansion and growth plans. This presentation was a natural segue into his second presentation, announcing the launch of the new high performance team.

"An important part of my expansion plan is the creation of a new high performance team that will report directly to me. The members of this team are Ian Foster, regional director of Western region, Bill Collier, branch manager, Keswick, and Barry Wilson, branch manager, Chicago. Will the three of you please stand up?"

The three stood up and there was a round of applause.

"They will remain in their current jobs and locations but will assume additional responsibilities for company growth. In addition to their current jobs, they will look for new opportunities, new markets, and come up with ideas on how to grow existing markets."

That last sentence jarred me. Wasn't growing existing markets a marketing responsibility? Wasn't finding new markets a marketing responsibility? In my Maritime marketing plan, there was a dedicated section on new market opportunities, plus a proposed strategy that outlined in detail how we should go after those markets. What was Don up to? In his talk, he didn't once mention how these three were going to interface with marketing. They were reporting directly to Don and dealing only with Don. This was strange.

Don's last announcement had left me deflated, with the wind blown out of my sails. It left so many unanswered questions in my mind. As I looked into the future, I saw conflicts and turf wars. Is this what Don was intentionally setting up?

Despite my misgivings about the high performance team, I was excited and looking forward to doing my presentation. I had to get myself together. The 10:45 coffee break was coming up, and I was speaking afterwards. I had to focus on the present. We had a solid marketing program. We were going to achieve great things with this program, and there was no point in worrying about the future. I had to take one day at a time, and now was my time.

My presentation was well received. When I finished, the applause was loud and long. I glanced over at Don. He was smiling and had a look of satisfaction. I was happy.

At lunch, I was approached by many of the regional representatives, expressing their enthusiasm for the marketing programs. "It's nice to see something different for a change." "It's about time we had a proper marketing program." "I can hardly wait until May. This is great."

The encouraging comments helped me forget about the high performance team. If we stayed the course and did a good job, we would be all right. The afternoon flew by. At the cocktail party, I met a lot of people from the regions I had been hoping to meet. All in all, I felt pretty positive at the end of the day as I drove to the hockey game.

I arrived at our box at 7:15. There was a good crowd on hand already. A buffet had been set up against the left wall, and there was a bar and bartender immediately to the right. I migrated to the bar first, where I saw Andy and George. Both of them seemed in good spirits. I was thirsty, so I ordered a beer.

"Hey, Michael, how are you feeling?"

"Great. How are you two?"

Andy spoke first. "Good, I'm feeling real good. I think we had a successful day."

"That went well today, Michael," George said with his infectious grin.

"Yeah, I thought so, too. The only thing I'm a little concerned about is this high performance team," I replied.

"Don't worry about it, Michael. It will all play out okay in the long run. Let's enjoy the moment." Andy seemed in a particularly good mood.

The three of us moved to the buffet table, where there were fresh slices of pizza and an assortment of deli sandwiches and wraps. There was the usual vegetable platter of carrots, celery, green peppers, and zucchini. I helped myself to a slice of pizza and a couple of wraps. I scanned the room to see who was there, or rather, who wasn't there. I expected all of the head office VPs and directors to be there. The box was large enough for a couple of dozen people. I noticed that the three members of the high performance team were congregating in one corner. Don was speaking to them. I also noticed that Craig Lofthouse was not present. I searched for Marcia Hanson and couldn't find her either. When I discovered that only some of the summit attendees had been invited, I wasn't surprised to see that none of the branch managers was present.

The game started promptly at 7:30. The Toronto Maple Leafs were hosting the New York Rangers. Some watched the game live from the two rows of seats that were outside the box. Others were engaged in discussion and glanced at the game on the TV monitor in the inner part of the box. There were still others who sat on the U-shaped sofa, chatting among themselves and disregarding the game completely. The remainder were gathered around the bar.

I really wanted to talk to Don about the high performance team. I scanned the room and watched for an opportunity to speak to him. As it happened, Don was working the room and he approached me.

"Did you have enough to eat, Michael?"

"Yes, thanks, Don. This is a very comfortable way to watch a hockey game."

"Yes, it's not bad, not bad at all. Looks like everybody liked your marketing programs. I've been hearing complimentary remarks from everyone I've spoken to. Most importantly, the field people are happy and they're looking forward to May 1st." There was a loud roar from the crowd. Someone must have scored, but I didn't pay any attention. I was focused on Don. I had my opening, "Don, do you have a minute to chat? I'd like to ask you a question."

"Sure, shoot."

"I wanted to ask how you saw the high performance team interfacing with marketing. Based on your description of their responsibilities, it looks like they're going to be doing similar things to marketing."

"I don't see that at all. You just keep doing what you're doing, and they'll do what they'll do under my personal direction. I see their scope as being much more strategic than yours."

"I understand. But, for example, I'm completing the Maritime marketing plan now. It will identify new market opportunities and provide a strategic direction. From the sounds of what you said this morning, the high performance team will be doing the same thing."

I could see Don's mood changing. He was starting to get annoyed. "Don't question what I do, Michael." His voice was slightly elevated and very firm. He was glaring at me. "I run this company my way, and I won't repeat this again. You do what you're supposed to do, and the high performance team will do what they'll do. Period!"

He walked away. There was another roar from the crowd. I realized I had just spent all the credits I'd earned today. I felt stupid and depressed. Why couldn't I leave well enough alone? Why couldn't I let things unfold and see what would happen next? I took a glass of red wine from the bar. I walked to the edge of the box and sat down in one of the outside seats. I watched the game, but I didn't see it. I didn't know the score. I really didn't know the score.

Wednesday & Thursday

After the summit ended, I went back to my office. I had to finish the Maritime marketing plan by Thursday evening. I'd promised Tom McAllister, the regional director for the Maritimes, that I would send him the sections dealing with the market analysis, new opportunities, and marketing strategy by Friday noon. I felt a sense of relief because Don had cancelled the Friday morning meeting after the summit. This would give me some breathing space. I rarely shut the door to my office, but that's exactly what I did Wednesday afternoon and all day Thursday. I asked Cheryl to take all my calls. I didn't want to be disturbed unless there was an emergency. She told me Don was not available to see the logo proposals for the next two weeks. He had wanted some of the vice presidents to be present, and after not being able to find a time when everyone was going to be available, the logo presentations were arranged for our Friday morning meeting in three weeks.

Week Eight

I was able to complete my draft plan by Friday morning and sent relevant pages to Tom before noon. I asked him to provide me with feedback by the end of the day on Tuesday, so that I could finalize the plan and have a few copies bound and in Don's hands by Friday afternoon. I noticed my eyesight was getting worse.

On Friday morning, I received calls from all the regional managers. In light of the new marketing programs, they wanted to know if they should go ahead with the sales programs that were in the marketing playbook for March. The March program offered 10 percent off cooking stoves. Like all of the marketing playbook programs, this one had a short shelf life of one month. It comprised small ads offering 10 percent off any stove in our stores for March. The ads were designed to run in small local newspapers for the first two weeks of the month. It seemed reasonable to respond in the affirmative. Nevertheless, I told them to hold off until I'd had a chance to speak with Andy.

At 10:00, when I went to Andy's office to talk about the March stove sale, I discovered that he and Don had left for Chicago early that morning. They were not expected back until Thursday, late afternoon. The regional directors wanted an answer by Tuesday, so that their branch managers could get the ads in for the beginning of the week. Our program was two months away from launch. Cheryl couldn't get Andy's phone number, which I thought was strange. Why couldn't they be reached? I called the regional directors and told them to go ahead with the March sale.

The rest of the week was peaceful. There were several meetings to review and discuss various aspects of the marketing programs. There were no major problems, and the work was proceeding as planned. I received feedback from Tom on the Maritime marketing plan. He made a few good suggestions and they were incorporated. Cheryl made four copies of the plan and had them bound.

Friday

I thought we'd learn something about the mysterious trip that Don and Andy had taken to Chicago, but nothing was mentioned at our Friday morning meeting.

"We have a lengthy agenda this morning, so let's move forward. Dick, you're on deck first with the maintenance program."

Dick Howard was the director of maintenance, reporting to Stan Ward, VP of operations. He had a sizable portfolio that included the maintenance of all of our trucks and equipment across the country. He was originally a mechanic who'd worked his way up the corporate ladder over a period of thirty years to his present position. He'd had this position for the past two years. Dick looked uncomfortable in his shirt and tie, and he wore a brown suit that didn't fit properly. It looked like he'd bought it when he was twenty-five pounds lighter.

Dick presented his plan. The final plan had to be approved in March so that it could be activated in the spring. The maintenance work accelerated through the spring into summer, when the bulk of the work was completed. This year's plan was a departure from those in the past in that this was the first year Northern Energy was exploring the possibility of contracting out a sizable portion of the work to local mechanics. Dick was visibly nervous as he presented his plan. There were a lot of questions. But Dick fielded most of them satisfactorily, and those that he needed help with were answered by Stan Ward.

Don made a few suggestions. "When you've incorporated the feedback from this morning into your plan, bring it back here next Friday and we'll have another look at it. Okay. Larry, there are some technical issues to resolve. You're next."

Larry Field was the director of technical services. He also reported to Stan Ward. He was responsible for technical support, inspections, and safety. A year didn't go by without an explosion or fire somewhere. Larry had been on the job for only a year. This was a high-profile position because of the potentially hazardous nature of our business. We could not afford negative publicity, because it immediately impacted future growth prospects. The turnover rate in Larry's position at Northern Energy was high.

"First, we're right on schedule with our tank inspections. We haven't found anything out of the ordinary that we couldn't fix within the parameters of our budget. Our valve replacement program slipped a little, because we experienced a greater number of regalements than was originally anticipated. The lost-time injury rate is tracking about the same as last year, and the programs we put in place have not had a chance to have full effect."

"Bullshit!" Don shot out of his chair as if someone had placed a bomb under it. His faced turned a crimson red. "What the fuck is going on? I heard that in Sackville last week, a technician had his eyebrows singed right off and was lucky he didn't have the hair on his head burned off, because he wasn't wearing all the right protective gear. Shit! What is the matter with you people? Marcia, what's the status of the certification of our new field technicians? Where are we at with the safety training? What do we have to do to improve our lost-time-to-injury numbers? I've had it with you morons." He was shouting at the top of his voice, and the level had not changed. "If you assholes cannot get your shit together by next week, don't bother coming into work. I don't want to see your stupid faces again. I don't want any more fires! I don't want any more injuries! Do you hear me?"

Larry and Marcia just sat there in silence.

"Well, do you hear me, you clowns? Are your ears filled with wax?"

"Yes, I hear you," Marcia and Larry said in unison.

That was the last tirade of the meeting. We finished shortly before 11:00.

I returned to my office, and a few minutes later, Cheryl told me that Marcia had run out of the building in tears. She'd quit.

Week Nine

I felt anxious all weekend. I knew something was wrong. At 8:30 Monday morning, Don called me. "Michael, I don't get this Maritime marketing plan. I read it over the weekend. I didn't get it, so I read it a second time. I thought maybe I missed something. After reading it the second time, I still don't understand what you're trying to get at. It's not acceptable. You'll have to do it over again." He was perturbed, but he didn't raise his voice.

"Don, is it the strategy that you didn't understand, or the recommendations, or what, specifically?" I needed information to build on if I was to rework the plan. How much did I have to rework?

"Everything," said Don. "There's a lot of information, but I don't like any of it. So redo it."

That was the end of the conversation. I wished I could take a look at the Western marketing plan that had been done for him by a consultant in Calgary, but which I was forbidden to see. No one in the marketing department had seen it, so I couldn't glean any information from them. I couldn't understand what Don objected to. I had followed the typical structure of a marketing plan that I was taught at the Columbia Graduate School of Business in New York, and at the Wharton School of Business at the University of Pennsylvania. It was acceptable at Ontario Hydro and at Avstar Aerospace. Yet it was not acceptable at Northern Energy.

I called Andy. Luckily he was in his office. "I just spoke to Don and he asked me to redo the Maritime marketing plan, but he wasn't any help explaining exactly what he didn't like. Have you read your copy? Can we talk about it for a few minutes?"

"Yes, we can. Don already told me he didn't like it. Why don't you pop over now?"

I picked up my copy of the plan and walked to Andy's office. "Did you not understand it, Andy?"

"I had no problem with it, Michael. Where you missed the mark is that you forgot where Don's coming from. He's a finance guy, a chartered accountant. That means he's meat and potatoes. You have a section on demographics and the changes in the demographic patterns that are taking place in the Maritimes. That's fluff to Don."

"But we're in a retail business," I protested. "Changes in demographics are key to developing the strategy that's contained in the plan."

"I understand all that, but that's not how Don sees things. He shapes people and events to fit his view of the world. I think that's why he thinks a lot of the information and analysis in the plan is fluff. His approach is: I've got product to sell. Now lets get out there and sell the product. That's why he likes your new marketing program. On the one hand, it's fancy because of all the bells and whistles. But on the other hand, it's just a dressed-up plate of meat and potatoes. He understands it because it's highly focused. It's attacking a market and it has black and white elements. I think where he's getting lost in your plan is that you offer a 'positioning strategy' and a strategic plan. He wants programs. How much, when, what will it cost, etc."

"So he really doesn't want a marketing plan or a strategic marketing plan? He just wants a litany of tactics—in other words, programs?" I looked at Andy in total frustration.

"You hit the nail on the head. Give him tactics—programs—don't give him strategy."

"So what's the point of a marketing plan, anyway?" I asked. "In two months, we're going to be rolling out a comprehensive residential marketing program across the country, including the Maritimes. It'll be interesting to see the differences in the uptake from one region to the next. That experience will provide valuable marketing research into regional differences. But until we see what the uptake is by region, I'm at a loss as to what I can provide in the meantime. When I was asked to prepare the Maritime marketing plan, I was not given a budget to conduct original market research. I had to go with published data that allowed me to do market segmentation and produce a strategic direction. That's the best I can do with the information I have.

"Incidentally," I asked Andy, "what did you think of the positioning strategy?"

"I thought it was good, and I also think the markets you identified as up and coming, based on shifts in demographics and what is actually happening on the ground with energy use, were good. But it's not what Don wants. Develop some creative programs along the same lines as the one you're working on launching now, and that should satisfy him."

I got up to leave. "Thanks for your help, Andy. Do you have any idea how soon Don wants the changes? It didn't come up in our conversation"

"I wouldn't sit on it too long. Try to give him something in two weeks."

I left Andy's office completely frustrated. My plate was full and I didn't need this. I really didn't know what to do next.

Week Ten

Monday

I'd lost another couple of pounds. I was now 175 pounds, the same weight that I was when I was running eight miles each morning. My eyesight was getting worse. Now, straight lines were completely jagged. They looked exactly like the teeth of a saw. I also noticed that I was having difficulty reading letters clearly. I could still make out the words as long as the font was a twelve, but I couldn't read anything smaller.

Most of the concept artwork for the marketing programs was complete. Some of the pieces were at the printers, while others were going through our rigorous review process. There's nothing worse than finding a typo after receiving a hundred thousand copies from your printer.

Work had also begun on the playbooks that described the elements of the programs in simple terms and gave the field step-by-step instructions for the program. It also contained a Q&A section for as many questions as we could anticipate.

On the transportation trade show, we were able to get commitments from suppliers, private sector technicians, and large fleet operators. To our collective delight, we got top-notch speakers, including the deputy minister of the environment, who agreed to be the kick-off speaker.

I ignored the Maritime marketing plan the whole week. I just couldn't face it.

Friday

To my surprise, the Friday morning meeting went smoothly. Don was civil and restrained the whole time.

The three logo presentations were arranged for 9:00, 9:30, and 10:00. Fred Thompson was first, followed by Bob Marshall, and Peter Brown.

The three presentations were very different from each other. Fred chose the theme of our energy form as an environmentally desirable propellant. He combined two aspects—the energy itself and the environment—into one theme. Bob's logo renditions reflected our energy as a safe offering. Peter built on the theme of the versatility of our energy.

At the conclusion of the three presentations, Don asked what we thought. People weren't dying to jump in. Since this was my sponsored item on the agenda, I felt I needed to break the ice.

"I thought Fred's work was superior to the other two, both from a thematic and artistic perspective. I liked the way he blended together the two themes of energy and environment. His artwork was head and shoulders above the other two."

"Thanks, Michael. Does anyone else have an opinion?" Don looked around the room.

"I agree with Michael. I liked the look of Fred's the best," George said carefully, as if he were walking on eggs. No one else spoke up.

"I like Peter Brown's the best," said Don. "I like the color combination of the logo, and I like the fact that Peter prepared more alternatives to choose from than the other two." Don had a look of satisfaction on his face. It was clear he was going to choose the logo he preferred. It was also clear that everyone knew it, and that was why they hadn't voiced an opinion.

I was relieved when Don concluded the meeting by saying we needed more time to consider the designs. "Michael, ask each of the firms to submit their invoices for $1,000. We'll need more time to consider this. The implications of a new logo are huge, and we can't rush into it. But I do prefer Peter's proposed logos."

Week Eleven

The high performance team that Don had created the previous month was beginning to show its ugly head. A couple of weeks earlier, there'd been a request from Bill Collier for some general marketing data for the commercial sector. Last week, there'd been a request from Barry Wilson for more detailed information for the industrial sector. This week Ian Foster was asking for similar information for the transportation sector. Naturally, all of these requests came straight to my office.

I didn't pay too much attention when I received the first request. I passed it on, figuring it wouldn't take long for the commercial marketing specialist to gather the information. When I received the second request, I saw it as an irritant. But after the third request, my blood began to boil. I could see the wedge firmly establishing itself in the marketing department's door. In addition, we couldn't afford to add workload to the teams that were preparing the two new marketing programs. There was too much for them to do as it was. They needed to stay focused, not spend time chasing information for unrelated work.

I called Ian. He wasn't there so I left a voice message asking him to call me. When he didn't return my call, I sat on his request.

Two days passed and still no call from Ian. I picked up the phone and called him again. No answer. I left another voice message. I knew he had call display, but I didn't know if he was avoiding me. Because he was in the West, I couldn't just walk to his office to see if he was there. I called his EA, and she informed me he was in a meeting. I asked her to get him to call me.

I was also annoyed by these requests because I wasn't told why they needed the information. I had no idea what Don had them working on. And I was not going to ask Don.

On Wednesday, Don called. "Michael, I got a call from Sam Schwartz, and he told me you haven't paid his invoice."

"That's correct. In January, Sam promised to courier a copy of the December issue of *Truck and Trailer* so I could see what the ad looked like. He never did. We have no record of this publication in the office. George Cornell has never seen it."

Don interrupted. "I don't care if George has seen or not seen the magazine. Just pay the invoice."

"But Don, I spoke with Sam a few days ago, and he promised he'd send me both the December and January issues, and I still haven't received them."

"He told me he sent them to you. Pay the invoice or I'll fire you. Case closed. Incidentally, where is the revised Maritime marketing plan?"

"I'm working on it, Don. When would you like it?"

"How about yesterday?"

"I'll have the revised plan completed in one week."

"Why the hell is it taking you so long?" He hung up.

"Oh shit" was all I could say to myself. I didn't know how much more of this I could take. I was finding it harder and harder to get out of bed in the morning. My energy was low, and I know I was lousy company at home. I didn't spend much time with the kids anymore on the weekends, and I hadn't gone out with Debby since October. I had to work on the weekends to get the marketing plan completed. There just wasn't enough time during the week to dedicate to it. Plus, I had completely lost interest in sex. Several times in the last couple of weeks, my wife had asked me if I was okay. I'd sort of dismissed her question, saying, "Yeah, I'm just a little tired. I'll be back to normal after May 1st when we launch the first marketing program."

I signed the *Truck and Tractor* invoice and passed it to Cheryl for processing. It was for $14,400 and covered the first six months.

Friday morning arrived and Ian had not called, so I decided to send him a note: "I received your request for the marketing data for the transportation sector. When do you need the information? What's it for? Please, let me know."

At the senior management meeting, everyone was present except Craig Lofthouse.

"Good morning," said Don, looking around the table. "Before we get into our agenda this morning, I'd like to announce a reorganization of the company."

My first thought was: Not another reorganization. It hasn't been a month since he announced the introduction of the high performance team.

"I want the company to be more agile, more responsive, so that decisions can be made more quickly. To do that, I need to streamline the organization. I'm combining merchandising and marketing under one vice president. That VP will be George Cornell. As a part of this restructuring, Andy will move to Chicago where he'll head up U.S. operations. Craig Lofthouse has left the company."

It was a nice way of saying that Craig had been fired.

Week Twelve

Only four weeks to launch for the residential program and six weeks for the transportation program. Although I was tired, I was also excited because all the program pieces were coming together beautifully. We got the permits for the contests that were an integral part of both programs. The marketing material looked spectacular. The "Sizzle into Summer" program was coming to life. The signage was complete and in production. To keep everything as simple as possible so that we could meet our tight timelines, the indoor and outdoor signage was made of the same material—chloroplast. Although this added to the cost, it provided a more durable and more professional-looking sign. If we chose to run the program again next year, the signs could be reused.

The teams had started working on the program instruction manuals. They were arranging trips to the regions to provide hands-on instructions on the residential program. There was no time to go to each branch. Five regional meetings were arranged. The branch managers and their sales staff would attend these meetings. I was hoping to make it to one or two, but at this stage, I wasn't able to commit.

The organization for the transportation trade show was also coming along well. The venue was reserved. All the speakers had been lined up. Seventy percent of the booths were booked and two sets of event brochures were ready. The first set was designed to build awareness for the event with the event look and theme—Propelling You into the Future—a few punchy words, and the date. A final event brochure was completed when all the program pieces were finalized. Since we already had the artwork from the first mailing, the turnaround on the final brochure was fast.

On Wednesday, I submitted the revised Maritime marketing plan. I didn't like it. The strategic portions were removed, and some of the demographic trends were removed. I added a section on national programs (i.e., a repeat of what we were currently doing) and a section on customized regional programs.

On Friday, Don told me he liked the new plan, and then he proceeded to berate me for not cooperating with the high performance team. What was going on here? I wondered. Was this some sort of elaborate set-up? What was the real reason Don had established the high performance team, and what was their real purpose? Was I just being paranoid?

Week Thirteen

I shed two more pounds and leveled off at 173. I'd lost twenty-five pounds in three months. The morning workouts had paid off. My eyes were a little worse, but I didn't have time to see my doctor. There was no point in seeing him, anyway. He couldn't help me.

On Monday morning, Don walked into my office. "Michael, I want you to prepare a national hot-water-heater program, and I want it launched no later than Friday, and preferably by Wednesday."

"Pardon me?" What did he just say?

"Why do I have to repeat myself? You heard me. I want a hot-water-heater program launched this week."

I was stunned. I couldn't believe what I was hearing. I breathed deeply and spoke slowly. "Don, we're four weeks away from launching the residential marketing program. Water heating is one of the key components of the program. If we launch a separate hot-water-heater program now, just before launching the big program, we're just going to confuse our customers. In fact, we run the risk of pissing them off." I was shocked at hearing my own last words escape from my mouth. But I didn't care anymore.

"Michael—just do it, or I'll fire you and find someone else who will." His voice went up a few decibels, and his faced started getting red.

I sighed, "Okay, Don. We'll have something for you to look at in a couple of days."

He walked out of the office. I was still shaking my head. Maybe I should see my doctor, I thought. I wanted a Valium. I wanted a scotch. I wanted something to calm my rage and frustration. I couldn't believe the stupidity of Don's request. It was completely unreasonable. So, why should that surprise me?

I called Joe. His reaction was worse than mine. Despite all the pressure we were under, Joe always took everything calmly and in stride. He always maintained a serene air. This time, he blew his top. No more Mr. Nice Guy.

I shared an idea with him, and I asked him to help me flesh it out. Working on it seemed to calm him down a bit. By noon, we'd prepared a one-page flyer on 8 ½ by 11 paper, in a fancy font. It was a 15-percent-off sale for the month of April that the branches could offer. After showing it to Don, we would have Cheryl fax it to all the branches. But Don was nowhere to be found. I was tied up in meetings for the next two days, and then he was out of the office until Thursday at noon. I was finally able to see him with the proposed program over lunch. His reaction was almost dismissive. He really didn't seem to care.

"Yeah, fine. Goodbye."

Weeks Fourteen to Sixteen

The next three weeks were filled with positive energy and anticipation. All the residential program pieces came together. The program was magnificent. We split the residential team into two groups. One group went to the two Western region meetings and northern Ontario. The other group went the Maritimes, Quebec, and southern Ontario. Joe and I did Chicago and central and eastern Ontario.

The official launch of the residential marketing program went without a hitch. All the signage was up. Every branch had enough coupon books. The coupon books were stuffed into the April invoice envelopes of our current customers. These were mailed out the 21st of April to all our independent contractors. The branches prepared schedules to meet personally with all these contractors during the month of May. We had completed our work at head office. Now it was up to the branches to do their part.

Two weeks later, our "Propelling You into the Future" trade show began. The attendance was much higher than we had anticipated. We got positive feedback from the participants for the morning speaking/teaching sessions, as well as the afternoon exhibit and demonstration sessions. I was overwhelmed and gratified. We'd worked like dogs and pulled it off. The trade show was on Wednesday. Thursday evening, I joined the transportation team at the Keg for dinner. If ever there was a time to celebrate, this was it.

The residential team and I talked about going out for a celebratory dinner as well, but we'd agreed to wait until the end of May. We wanted to wait and see what the uptake to the program would be. By the end of May, appliance sales were up 18 percent over the previous year in May, and barbecue sales were up 30 percent. Early indications were that the "Sizzle into Summer" program was exceeding our projections. We were now ready to celebrate, and celebrate we did. We had a lot of steam to blow off.

The sales numbers were much stronger in June. In fact, they were so strong that the branches couldn't keep up with deliveries. The technicians couldn't keep up with new customer hook-ups and installations. By the beginning of July, the technicians were booking appointments for the middle of August, and in some branches, the beginning of September. Our whole sales and service system was straining. The program was planned to run to the end of July, but Don told us to end the program on July 10th, three weeks ahead of schedule.

I began to track the weekly sales numbers by product as they arrived on my desk every Monday morning. I looked at the top-line numbers to get an overall picture of where we stood. After that, I looked at the regional and branch sales by product category. Were there any differences? If so, what were they? Were there significant trends within regions or between regions that would help us better prepare for our marketing programs next year? The sales numbers from this program were going to provide us with excellent marketing information.

On Wednesday morning, July 17th, I got a call from Mary. "Hi, Michael. Can you pop up here for a few minutes? Don wants to see you in meeting room C."

"Okay," I said. "See you in a minute."

I walked up to the fifth floor. I was still high from the success of both marketing programs. They had exceeded all our targets by a significant margin. When I arrived at Mary's desk, I realized I had never been to meeting room C, which was right next to Don's office.

"Go right in, Michael," Mary motioned with her hand.

I entered the room seeing familiar faces. Don was sitting at the head of the table, immediately to the right of the entrance to the room. Peter, the corporate lawyer, was sitting at the other end. George Cornell sat to Peter's right. What puzzled me was Malcolm's presence. Malcolm Hubbard was a big, burly man. He stood about six feet, four inches tall, and I estimated he weighed between 260 and 270 pounds. He was not a man to be trifled with. He was standing in the corner of the room, behind and to the right of Don. Malcolm performed various jobs around head office, but his most visible job was as Don's chauffeur. I was asked to sit down in the empty chair immediately in front of me, to the left of Don and opposite George.

This was a small meeting room. It was stark in comparison to Don's office and to the boardroom. It didn't fit the motif of the fifth floor.

Don started the meeting. "Michael, you're a bright guy and you really know your stuff. The marketing program was a success. But I'm going to have to let you go. There will be no discussion about who said what and when. Your termination will take effect immediately. You will receive full salary to the end of the year. However, you will not get benefits and you will return the company credit card. Peter will leave here with you. He will accompany you to your office to collect your personal belongings. Then he will escort you to your car. May I remind you that you signed a non-compete agreement when you joined Northern Energy? The non-compete agreement forbids you to work for another energy company for two years. If you do, I'll sue your ass."

The whole scene was a dream I was watching in slow motion, like an out-of-body experience. I asked, "Don, where did I let you down? How have I not met your expectations or objectives?"

"I said there would be no discussion," Don replied. "You heard the terms of separation. I'm dealing with you generously. Goodbye," he said in an even and calm voice. Don got up to leave. Malcolm followed right behind. Now I knew why Malcolm was in the room. Don was about five foot six and medium build. Malcolm was Don's bodyguard as well as his chauffeur. It made sense. Firing staff was a full-time job for Don and he didn't leave anything to chance—physically or legally. I could not sue for wrongful dismissal. He was paying me full salary to the end of the year. I had only worked for six months (even though it felt like a year) and he was giving me six months' severance pay.

Peter came over to me as I stood up. I still thought I was dreaming. This was not really happening. It couldn't be happening. Peter escorted me to my office where I picked up my personal belongings. I didn't have many, just my gym bag and a couple of family pictures. There were some other things I would have loved to take: a copy of the Maritime marketing plan, a copy of the "Sizzle into Summer" coupon book, and other associated program pieces. The phone rang. I looked at Peter and he nodded. I picked up the phone. It was George Cornell.

"Michael, can you come up to my office for a second and bring your copy of the Maritime marketing plan?"

I looked at Peter. "George wants me to come up to his office for a second."

"Sure, let's go."

We went back upstairs. As I walked toward George's office, I saw Don out of the corner of my eye. He was leaning over a cubicle partition talking to one of George's staff. I'll never forget the look on his face when our eyes met. He was slightly flushed. He looked pleased with himself, as satisfied as if he'd just finished a steak dinner and had his fill. Firing his employees gave Don sustenance. It fed him.

I forget what George wanted. It didn't really matter. Peter and I went straight from George's office to his office. I took the company credit card from my wallet and gave it to Peter. I also gave him my ID pass for the security locks on the doors. He gave me another copy of the non-compete agreement.

Peter walked with me to the front door and watched me walk to my car. No more than ten minutes had elapsed from the time I'd received Mary's telephone call. I was in complete shock. Everything had happened so fast. It had been a surgical cut and, like all such cuts, you don't feel anything when the first incision is made. There's a delay before the pain comes. I sat in my car in the parking lot. I couldn't drive. I was suddenly paralyzed with pain. What did I do wrong? Not knowing why I was fired was worse than being fired. I knew Don was a madman. But I still wanted to know why. I needed a reason. I felt emotionally and physically violated.

I turned the ignition key. While I drove, my mind wandered to my family. What was I going to say to my wife? I didn't want to inflict pain on my family, but it was exactly what I was going to have to do. How was Debby going to react? How were the kids going to react?

Fast Forward a Few Months

I took time off to lick my wounds. I started to reconnect with former business colleagues. I noticed in September that my eyes were getting back to normal. Straight lines started to look straight again. However, by November, I was emotionally spent. I didn't want to get out of bed in the mornings. I was sad most of the time. I had no energy, no motivation. I couldn't face the prospect of looking for another job. I was falling into a dark abyss. Debby and I talked about it. A good friend of ours in the neighborhood was a psychiatrist. She suggested I make an appointment to see him, which I did. I was diagnosed as clinically depressed and was under his care for the next two years.

Follow-up Events

I kept in touch with Bob Marshall, president of Campbell Marshall. Bob was very supportive in my search for another job. He placed the resources of his office at my disposal. They helped me with graphic design and print material for my new resume. We kept in touch for several years, and Bob was also my conduit to the news at Northern Energy.

The marketing program was the most successful in the history of Northern Energy. Year over year, sales of all hard goods increased by 37 percent. Energy sales increased by 7 percent. Don terminated Peter, the corporate lawyer, in October. The high performance team was disbanded in November as part of another reorganization. In April the following year, Don was summoned to a private meeting with the chairman of the board. The numerous hirings and firings of managers and staff at Northern Energy had been brought to the attention of the chairman and the members of the board. Apparently, a quiet investigation was initiated. The board received a report that documented the turnover rate of both staff and managers for the five-year period marking Don's tenure as president. Northern Energy was one of many companies that were part of a large conglomerate. The turnover rate of staff at Northern Energy was four times higher than the other companies in the group. The turnover rate for managers (this included managers at all levels) was twelve times higher than the other companies.

Apparently Don's firing was swift. There was no discussion. Don walked into the chairman's office, and the chairman told Don he was finished. He was escorted out of the office by two security guards. It seemed like a fitting end.

I kept in touch with a consultant who periodically worked for Northern. He informed me that the following year, Don was hired by a medium-size electric utility. In his first week of work, he terminated half a dozen senior managers. There was a huge political backlash. The board was criticized by politicians for hiring Don, and there was a great deal of negative press. Two weeks later, the board terminated Don's employment. Lawsuits followed. Don sued for wrongful dismissal and the utility counter-sued for misrepresentation. I lost interest. I didn't want to know anymore. However, I later discovered that Don had started a one-man, one-truck energy business. Apparently, he had no desire to grow beyond the one truck because he was tired of being surrounded by idiots.

Reflections

The head office of Northern Energy was beautiful, a showpiece. It was clean and tidy, almost sterile. There was no life, no humanity, and no spirit. Everyone kept their head down, maintained a low profile, stayed out of sight, and just did what they were told. Creativity and individual initiative were unknown quantities. You did what you were told or you were fired. You checked your emotions, personality, values, integrity, and humanity at the front door before you entered the building. It was a dictatorship ruled by fear.

I don't know what the personal toll was on the employees who worked at Northern for any length of time. I only know the price I paid. I only know how I reacted physically and emotionally—first with my loss of sight, followed by clinical depression. Firing staff was a ritual for Don Edwin. It was an expression of his power, and that power was intoxicating. It gave him a high, and he had to satisfy that high on a regular basis, just as an addict needs a regular hit. Northern was a revolving door for employees, especially at the senior management level. How were all those individuals affected when they were fired?

When an organization is ruled by fear, mistakes are made, and hiding mistakes is an essential part of survival. The time used to fix and hide mistakes reduces productivity. This was evident in the field as well as at head office. Some of the more serious mistakes could not be hidden. There was an explosion at a customer site during my tenure at Northern when heads rolled.

Beyond the human toll, one wonders how Northern's bottom line was impacted by Don's management style. How much better could the bottom line have been if new ideas had been allowed to percolate up the line, instead of being forced down the line?

Comments by Dr. Don Fulgosi: A Medical Perspective

Don exhibited typical characteristics of a psychopath. There was a total absence of compassion or empathy toward his employees. He was ruthless and derived sadistic pleasure from abusing, bullying, threatening, and firing his employees. He experienced no feelings of guilt or remorse for his actions.

He saw relationships in terms of dominance and submission. He insisted that everyone submit to his way of doing things. It was his way or the highway. He had a huge ego. He exerted his control by micro managing. All of these characteristics created a work environment of huge "effort—reward imbalance" issues, with unfairness, injustice, uncertainty, unpredictability, and helplessness controlling the atmosphere. Helplessness—in humans, as well as in rats—is another prescription for disease and death.

Part 2

Research Results:

Connecting workplace practices and leadership behaviors with employee engagement and employee health

Chapter 1

Organizational Alignment: Developing a Diagnostic Tool

After Northern Energy, I worked for a private holding company with six separate companies in its portfolio. This company was in the waste disposal and energy markets. The waste disposal portion of the business had a twist. It collected only waste wood. It manufactured wood-burning furnaces and pellets that are used as fuel for pellet fireplaces. It owned and operated a large steam plant that sold steam to local industries.

The steam plant had been built at the beginning of World War II to provide steam energy for a new industrial park that had been built specifically to support the war effort. The park housed various types of munitions factories (explosives, cannon shells, bullets), textile and uniform manufacturing, and other companies manufacturing products for the armed forces. After the war, many of the factories stayed and refocused their production on commercial products. For example, DuPont and Dominion Colour remained. Ajax Textile remained and refocused its market from dyeing cloth for the military to dyeing cloth for all types of fashion apparel. Another large user of steam was a manufacturer of instant coffee. New companies that needed large quantities of inexpensive steam for their manufacturing processes moved into the industrial park. These included Styrofoam manufacturing and companies that made various kinds of foam products. At the time, the steam plant was owned and operated by the town of Ajax.

The town was losing money, and in 1985, it decided to get out of the steam business. It sold the plant to our company for one dollar. The private company converted the plant from burning oil and natural gas to burning waste wood. Natural gas was kept only for backup purposes. The "tipping fee" (disposal fee) for waste wood was $150/ton during the mid-eighties and early nineties. After the conversion of the plant to wood fuel, our company enjoyed immediate positive cash flow. It received income for its fuel to make steam and then sold the steam to local manufacturers at a rate that was a little below the cost that these manufacturers would pay to produce their own steam. It was a good deal for the customers and a great deal for the steam plant.

The capital cost of converting the plant to burning wood was quickly recovered. An established and captive market was in place. The steam distribution lines had been built during World War II. Some of them were getting old and developed the odd leak, but this wasn't much of a concern. As long as the leaks didn't present a safety hazard, the extra wood that was being burned was income and not an expense. This was the thinking before I was hired. The steam plant was a moneymaking machine.

In 1991, I was hired by the holding company as vice president, sales and marketing, responsible for the marketing and sales activities for all six companies in the group. The other companies were all related in some way to wood energy. There was a company in Manitoba that manufactured wood-burning furnaces, a plant in Montreal that manufactured wood pellets for pellet stoves, a waste transfer station in Brampton (on the west side of Toronto) for collecting waste wood, a construction company, and a waste management company.

It was a family-owned-and-operated business. The owner was my age. He had a son studying business at a university in Ohio and a daughter studying general arts in Toronto. Each summer, the son and daughter would work in the business. I was hired because the business had grown too unwieldy for the owner to run, even though he had managers who ran the day-to-day operations of each of the units. He had just purchased the companies in Montreal and Manitoba because he knew the markets for their products were substantial. He also recognized that he now had to grow these companies, and he needed the expertise to make this happen.

It was also apparent that the company was run by an entrepreneur who did what he felt like doing. As a result, the company began to lack strategic direction. With the purchase of the two new companies, the financial controls and management systems were inadequate. Up to this point, there was no formal marketing or sales function. Also, in 1991, the tipping fees began to fall. They dropped from $150/ton in 1985 to $100/ton in 1989, and were continuing to slide when I arrived.

After some preliminary discussions with Gord, the owner, on the major issues facing the company, I quickly completed an operational audit of the three companies that were at highest risk: the steam plant, because of the rapidly falling tipping fees, and the two new companies, whose sales and revenues had to be significantly boosted. The audit findings were very interesting:

1. The only contact the company had with its steam customers was the monthly steam bill, and on the anniversary of the steam contracts, when a letter was sent to each customer advising them of increases in their steam rates. The steam revenues ranged anywhere from $250,000 per year for the smallest company, to several million dollars for the larger companies. Most of the companies fell into the later category.

2. The two new plants were inefficient and the cost of production was too high. The company in Montreal had a little sales effort, but the company in Manitoba relied strictly on word of mouth.

It's not the purpose of this book to describe the work that was done to boost sales, productivity, and revenues, but rather to talk about employee engagement. Nevertheless, here are a few highlights to provide a context:

1. I met with the president of each of the steam customers. In every case, I discovered areas of dissatisfaction, and I was able to follow up and deal with these problems. In the case of a hospital, I discovered that they were planning to drop us as a supplier, and buy and install their own steam boilers. I learned about their greatest areas of dissatisfaction and the reasons for wanting to switch. I was able to arrange a formal meeting that included the key decision makers of the hospital, Gord, and myself. At this meeting, we addressed all of their concerns and were able to convince them to continue with us.

2. In the case of the Montreal pellet plant, I found a man who was a leading expert in the pellet market. With a little persuasion and an attractive financial package, he agreed to join the company as sales manager. John had an intimate knowledge of the pellet market, and he also knew most of its largest purchasers. He was able to increase sales to the point where they met our manufacturing capacity within a year.

3. The wood-burning stove company required a great deal more work. I spent some time in Manitoba to work with our local manager, who was the previous owner, to redesign and expand the plant. This was a mom-and-pop operation when Gord bought it. Charlie had been producing about one stove each month. This was fine for Charlie, since each stove was sold for about $5,000. It provided Charlie with gross revenues of

$60,000, which was acceptable for rural Manitoba. Gord bought Charlie's company for the potential it offered. After one year, production grew to 60 units per year and $300,000 in gross revenues. It was still small, but we were headed in the right direction. I was able to sign up rural heating and plumbing contractors as distributors in northern parts of Quebec, Ontario, Manitoba, and Saskatchewan, and in the northern parts of the United States. In fact, our fastest growing markets turned out to be in Minnesota, North Dakota, and Montana, and to a lesser degree, Wisconsin and South Dakota.

After one year, Gord asked me to assume, in addition to my sales and marketing responsibilities, full operational responsibilities for our pellet plant in Montreal and our wood furnace plant in Manitoba.

With the great distances separating both of these companies, I needed to have better reporting systems than the ones that were in place at the time. I was receiving monthly financial statements, typically one month after the fact, which meant the financial data was already out of date. I was also receiving weekly sales data providing information that was current, but it only showed one side of the business equation. What I lacked was predictive data—information that would give me a heads-up on what might happen three or six months in the future.

Measuring Organizational Alignment

I began to dust off the value analysis profiles that Graham Tucker and I had developed fifteen years earlier to see if there was anything that could help me build a diagnostic that would be predictive in nature. During my review of this material, the company president was organizing a strategic planning retreat for the twelve senior managers of the company. He asked me to help him with the retreat. This motivated me to think about our entire business and some of the problems we were having. The more I thought about these problems, the more I began to realize that an effective way to assess the root cause or source of the problems was to understand the point of contact of all our business relationships. At the time, I was also reading a great deal about high performance companies. Frequently there was reference to "organizational alignment" as a key ingredient of a high performance company.

Stephen Covey writes, "*In the future, systems reengineering, excellent customer service and manufacturing defect-free products to specifications will not be enough. The future is now. The enduring, empowering high performance organization of the*

future will require total organizational alignment to foster innovation, nurture con-
tinuous improvement and sustain total quality results."

Specifically, organizational alignment is the extent to which all organizational components are working together in harmony, supporting each other, and moving in the same direction. Organizational alignment is essential for providing exceptional service and maximizing profits.

Covey continues, "*Total organizational alignment means that within the realities of the surrounding environment, all components of your organization, your mission, vision, values, strategy, individual styles and skills and especially the minds and hearts of your people, support and work together effectively for maximum performance.*"

In their concluding remarks in the book, *Built to Last,* James Collins and Jerry Porras write, "*Most important, you've got to then align the organization to preserve the core and stimulate progress. Far and away the biggest mistake managers make is ignoring the crucial importance of alignment.*"

In the Epilogue of *Built to Last,* in a brief discussion about IBM, Collins and Porras write, "*Finally we'd challenge them (IBM), to identify at least fifty more specific misalignments that inhibit progress. And then we'd challenge them to not just change these misalignments, but eliminate them entirely.*"

All the research demonstrated a compelling argument that total organizational alignment is an essential ingredient in creating a high performance company, which in turn means a highly successful organization. The evidence was clear. Companies that were highly aligned had above average profitability and were outperforming the stock market and their competitors. They outperformed their competitors by six times and outperformed the stock market by a factor of fifteen. They had highly motivated and productive employees. The companies were able to respond quickly to changes in the market and to broader changes in the social and economic environment. Customer loyalty increased. Their relationships with suppliers were very close, in many cases, seamless, resulting in high-quality products at a reasonable cost.

However, no one offered a methodology for measuring organizational alignment. Being able to measure alignment seemed to be taken for granted by all the authors, as one could see in the challenge presented to IBM.

When I began to think about measuring organizational alignment, I faced a number of challenges. In order to measure something, I needed to have a clear picture or definition of the object we were going to measure. A few guidelines were established. The picture or definition had to be comprehensive so that it could stand the test of time. It needed to have universal applicability so that it did not change as circumstances changed. The definition also needed to apply to different types of organizations in different sectors, with little or, preferably, no modification. This meant that the definition should be equally relevant and applicable to companies as well as government and not-for-profit organizations.

I love car racing, especially Formula One. It seemed to me that a car provided a perfect metaphor to demonstrate total organizational alignment. Everyone understands wheel alignment. When your car is out of alignment, you don't have as much control; the car is not as responsive, and the misalignment causes premature wear and tear on the tires. This results in increased operating costs, because the tires need to be replaced sooner than expected and the car uses more fuel.

If you extend the metaphor further to consider a highly tuned and complex racing car, total alignment becomes even more critical. If the car is equipped with an eight-hundred-horsepower engine, with a design speed of over 250 mph, that car better be equipped with tires, brakes, suspension, transmission, chassis, etc., that are in perfect harmony with the power of the engine. Taking it one step further, if the car is going to be driven competently to win races, it will need a well-trained and experienced driver who can extract the maximum performance from the car. Finally, for the car and driver to succeed, they will need to be supported by a well-trained pit crew, engineers, and designers.

Whether it's a Formula One, Indy, CART, or NASCAR racing team, it's designed to win. They share the same focus, purpose, and highly aligned structure to make them winners. All the components of their organizations work together in perfect harmony, supporting each other—man, technology, and machine—to achieve one clear goal: winning.

These identical principles apply to any business, corporation, or agency. Very simply put, organizational alignment means that all the complex components of the organization are working together and supporting each other in achieving a common vision and mission. It means that all the strategies, policies, processes, procedures, systems, leadership, values, management, and employee behaviors are completely synchronized and in perfect harmony to achieve peak performance.

In addition, racing teams don't operate in isolation. They are part of a larger external system—a league—with a governing body that sets standards for chassis design, engine displacement, and rules for racing. The league must deal with suppliers, lenders, investors, and shareholders. Teams are incorporated entities with an owner/president and a board of directors. They exist for the racing fans, or customers. The races must meet the fans' expectations for exciting entertainment. If they don't, viewership and revenues decline. Ferrari and Michael Schumacher dominated Formula One racing for the past seven years to such an extent that there was no more mystery about who was going to win. As a result, fan apathy began to set in. In 2004, the governing body of the Formula One league changed the standards in chassis design and lowered the engine displacement as a way of "leveling the playing field." Michael Schumacher didn't have a stellar year in 2005 but, as this book is being written, he is once again tied for first place in 2006.

All of the factors outlined above apply equally in business. A company has similar stakeholders to a racing team: employees, suppliers, lenders, investors, regulatory agencies, and a board of directors.

Depending on the sector, every organization has its own unique external relationships. When considering alignment, one must assess the impact of all organizational elements on these external stakeholders, as well as on the internal. How do all of the activities, systems, processes, procedures, behaviors, products, or services of the organization impact on these stakeholders? To what extent do they support each other and work together in achieving their corporate vision and mission as experienced by their stakeholders?

The Business Relationship Model

It became evident that alignment had a great deal to do with the relationship among all the elements that defined the business sphere of influence of an entire organization. In preparation for our strategic planning retreat, I developed the Business Relationship Model© (see below) to graphically depict these complex relationships. In subsequent years, after Entec Corporation was created, we developed a model for utilities and for healthcare to reflect the unique nuances of these respective sectors.

BUSINESS RELATIONSHIP MODEL

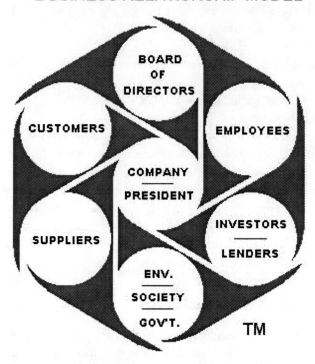

Having defined the "what" that we wanted to measure for total organizational alignment, the next problem was the "how." This problem was made considerably easier because the answer already lay in the model itself.

Since the model was based on relationships and the nature of relationships, the book *Service America*, by Karl Albrecht and Ron Zemke, came to mind. In their book, Albrecht and Zemke shared the story of the young man who took over the presidency of Scandinavian Airlines Systems, SAS. The president, Jan Carlzon, was given the mandate to turn the company around. Carlzon realized that the only way he could succeed was to make SAS totally customer focused. He recognized that as many as fifty thousand contacts were made between customers and SAS in any one day. He called these contacts "moments of truth," meaning that a judgment was made about SAS fifty thousand times each day. If Carlzon could guarantee that every "moment of truth" between SAS and the customer was positive, he would have a successful company. As we know, the rest is history.

It became evident that expanding the idea of moments of truth to all business relationships, both external and internal to an organization, rather than limiting them to customers alone, was a powerful way to measure organizational alignment. These moments of truth—or points of contact (POC)© experiences—are critical to defining the core of an organization, its values, vision, and mission. They define what the organization really is and what it stands for. It's at the POC that the organization becomes exposed. It's at the POC that the degree to which an organization is aligned becomes clear.

In simple terms, the Business Relationship Model is a behavioral- and knowledge-based model. The next task facing us was the development of the actual questions. As we considered each part of the Business Relationship Model, we decided to start at the center—the company. What questions were we going to ask? This clearly needed another layer of analysis.

We went back to our racing-car metaphor to find the answers. The car racing team contains all of the essential ingredients of an organization functioning in a highly competitive environment. It provides an excellent example of a high performance organization. The results are typically determined in two to three hours of racing. Success is measured simply—you win or you lose. The participants are totally exposed—there's nowhere to hide. There are no judges to assess and assign points. The team is totally accountable for its results. There are no excuses. There are only winners and losers. Some drivers and teams earn a great deal of money; some teams are money pits. They lose a great deal of money for their owners every year and are always scrambling to add sponsors so they can race for another year.

Developing the Questions

Our company became the perfect laboratory for developing a diagnostic tool that could be used to measure organizational alignment. Since Gord had asked me to develop an agenda for the workshop, I shared the model with him and the idea to use part of the time to develop the survey and test it among ourselves, as a way of gathering some important non-financial information about the company that we could use for planning purposes. Gord liked the model. But better still, he was supportive of this approach, and he was also enthusiastic about the prospects of gaining new insights into the company from the exercise.

The company had over two hundred and fifty employees and $50 million in annual gross revenues. It was geographically dispersed. Gord recognized that he did not know as much about the company as he would have liked. In preparation for our two-day strategic planning workshop, I sent each manager a copy of the Business Relationship Model and asked them to identify and list as many points of contact as possible from their vantage point. I also asked Gord to do the same from his perspective as the owner. I completed the task as well.

A week before the workshop, I received the information from all the managers. From it, I created a survey that contained all the point-of-contact information. The questions were organized under the seven sections of the model. When I had been the manager of marketing planning at Ontario Hydro, I'd had a budget of $3 million each year for market research. I learned a great deal about questionnaire design from the consultants who were awarded the market research contracts. This knowledge came in handy now. The questions were short and simply stated. The same five-point scoring algorithm was used throughout.

Twelve managers participated in the workshop and completed the survey. Even though this was a self-test, it revealed many interesting and profound insights. For example, Gord considered himself to be a man of high principles and morals. However, the survey revealed that the company treated many of its suppliers very badly. We were slow in paying our bills, and generally treated suppliers in a way that we would not want to be treated ourselves.

There were other significant leadership issues that emerged, identifying areas where Gord and a couple of other managers could improve. I had an excellent and collaborative working relationship with Gord, but some of the others did not. Gord treated everyone differently. He didn't provide managers with specific performance objectives. There were no annual performance reviews, and raises were completely arbitrary. There was no formal bonus structure. Business decisions were made on a whim from one day to the next, and no one knew what was expected of them. There was no company mission.

In many respects, it was a typical entrepreneurial company, where the owner seized upon opportunities as he saw them. He knew what he was doing, but no else did. He might consult some of us, but it was after he had already made up his mind.

Once, we ran into some problems with the pellet-manufacturing equipment in Montreal. When we bought the plant, we invested a great deal of money in improvements. The former owner was hired as the plant manager because he was considered an expert in pellet manufacturing. However, he made a serious mistake in judgment. He purchased a machine for manufacturing pellets from softwood when we were manufacturing pellets from hardwood. In fact, the demand for our pellets was high because our pellets were manufactured from waste sawdust that was sourced from the local furniture manufacturing plants north of

Montreal. We had a continuous supply of clean hardwood maple and oak saw-dust, which provided the best material for producing a premium quality pellet.

Unfortunately, the new machine kept breaking down, and we couldn't meet our commitments. In a marathon meeting with Gord and all the senior managers, we concluded that we should minimize our losses, close the Montreal operation, and move it to the steam plant, where we had enough waste wood to allow us to resume pellet manufacturing. The quality of the sawdust was not as good as in Montreal, but it was still higher than the sawdust used by many of our competitors, who used only softwood sawdust. At this meeting, we developed a detailed plan for moving the good equipment from Montreal to the steam plant outside Toronto. We even agreed on a fair severance package that we would offer the plant manager and other employees in Montreal. We owned several large tractor-trailers, and the drivers were told to be at the Brampton office first thing the next morning in preparation for going to Montreal.

The next morning, most of us were in the office by 7:30. Gord arrived at 8:30 and told us not to proceed as planned. He had woken up that morning and decided not to go ahead with the plant relocation. Time would show that this was a mistake.

The survey results showed that the company had reached a size and complexity where it needed more of a firm business plan and structure. We all knew this and the survey results brought it to Gord's attention. Since the survey was done in the context of our strategic planning workshop, it immediately provided the data we could use. During the workshop, we developed detailed action plans to resolve the many issues that had emerged. For example, process improvements were necessary at our pellet plant in Montreal. A specific plan was created to deal with this. Tipping fees had plummeted to $50 per ton. The old business model for the steam plant was no longer valid. At this rate, profits were being squeezed, and we had to look for and introduce efficiencies at the steam plant.

We were also sufficiently confident with the accuracy of the survey results that we could use it on a company-wide basis to gather further information from our employees. At the end of the workshop, Gord made a commitment to all the managers that he was going to work at making the changes in his own leadership style that had been identified by the survey. He committed to improve our relationships with suppliers. He also committed to support all of us in completing the action plans over the coming year.

The Quick Diagnostic Tool

Gord's son Russell, who was studying business at a university in Ohio, came to work for us in the summer prior to his last year at university. It was my first year with the company, and Gord asked Russell to work under my direct supervision. There was a great deal to do that summer, and Russell worked on a number of marketing and sales projects. When he left for his final year of school, Russell called me several times, asking for advice and input on major marketing or business papers he was working on. When he graduated the following year, Russell came into the business full time. Again, at Gord's request, he worked directly under my supervision.

Later that year, my wife and I went to Israel for a two-week vacation. A couple of things became clear to me on that trip. First, it was obvious that Gord was grooming Russell to run the business some day. This made perfect sense. I had turned fifty a month before the trip, and it also became clear to me that I wanted to go out on my own. I was energized by the process of creating positive change in organizations. I also became energized by the idea of offering the organizational alignment tool to other organizations as a diagnostic that would provide insightful information for a change process.

When we returned from Israel, I asked my wife how she felt about my leaving a permanent job and starting up my own company. Initially, she wasn't crazy about the idea, but the more we talked, the more it made sense. We owned our home, had no mortgage, no debts, and we had money in the bank. Plus, my wife had a full-time job. She agreed. I talked to Gord, and he supported me in my decision. Russ was settled in the company in his marketing role. My departure would give him the opportunity to assume greater responsibility. All the other elements were in place to continue to grow the business. I left, and on May 21, 1996, Entec Corporation was incorporated and launched. A colleague of mine, Dave Penny, a computer expert, provided the computer processing support. Dave and I decided to name the diagnostic the "Quick Diagnostic Tool" and we referred to it as the QDT©.

Before introducing the QDT© to the market, I developed a marketing plan. The key strategy of the plan was to approach the market on a segment-by-segment basis and target segments that I was familiar with and where I had contacts. I began with electric utilities. I had developed a close relationship with the general manager of a medium-size electric utility where the steam plant was located. I asked the GM if he would be willing to participate in a process where we could customize the QDT© for utilities. He agreed on a process that involved managers from every department in his utility. The managers started by providing their thoughts on the Business Relationship Model©. The model did not change in its overall construct but the content of the modules changed. For example, "board of directors" was changed to "mayor and elected councilors"; "investors and lenders" was changed to "regulators."

We then reviewed and amended questions in the modules that had not changed and developed new questions for the new modules. This resulted in the QDT for Utilities©, which was used by three utilities. Fortunately, these utilities were similar in size and had a similar customer profile (ratio of residential to business customers). This allowed us to compare the survey results to the financial performance of each utility. This correlation analysis showed that there was a direct link between the survey scores and the financial performance of the utility. It was so compelling that the results were presented at the American Water Works Association Annual Conference in Chicago in June 2000.

Soon after the QDT for Utilities© was completed, a similar process was followed for the healthcare sector and the QDT for Health Care© was completed. Despite the testimonials from the three utility general managers on the effectiveness of the QDT©, I was unable to sell it in the hospital sector. As I expanded the sales activities to other utilities and other sectors, there was a great deal of interest in the QDT©, but few sales. It became apparent that the QDT© was too comprehensive for the market. I discovered that few people are "systems thinkers," and if something is outside their box, they can't deal with it. I did, however, get interest in the individual modules, and especially the employee module. We were able to get several clients just for the employee module, and I started to refocus my sales efforts only on this module.

During my sales efforts in the healthcare sector, we made the decision to change the direction of Entec Corporation. I had arranged a meeting with the CEO of the Homewood Health Centre, a preeminent psychiatric facility with six hundred beds, specializing in treating substance abuse and depression. The CEO, Dr. Edgardo Pérez, had assembled a team of about a dozen managers in the boardroom. After the meeting, during the question and answer period, I received a cool reception from everyone except Dr. Pérez. He liked the way I was measuring alignment, but what he particularly liked was the employee module. We chatted privately after the presentation, and he told me about his work in the area of organizational health. He said that over 30 percent of the psychiatric patients at Homewood, especially those suffering from depression, were there because of their work and their organizations. He said he felt a personal frustration because he knew that after patients were treated, they were sent back to the same place that had spit them out in the first place.

A month before our meeting, Dr. Pérez and Bill Wilkerson had established the Centre for Organizational Health at the Homewood Health Centre. The announcement of the creation of the Centre coincided with the release of the book, *Mind Sets*, which they had coauthored. *Mind Sets* was a definitive work on the serious nature of depression, and especially the high cost of depression to business and to the overall economy. Bill Wilkerson was the former president of Liberty Health. Prior to helping Dr. Pérez establish the Centre for Organizational Health at Homewood, Wilkerson had established the Business and Economic Roundtable on Mental Health. The Roundtable comprised the CEOs of the major banks in Canada and the CEOs of some of its largest companies. Members also included prominent psychiatrists and other professionals who were associated with or involved in the area of mental health in the workplace. Dr. Pérez suggested that I meet Bill Wilkerson.

A week later, I met with Bill and he, too, liked the way we measured organizational alignment (i.e., the employee module). He said that he and Dr. Pérez were thinking of developing a diagnostic to measure organizational health and thought the Entec model was perfect, except that it was missing the mental health component. We subsequently discussed this with Dr. Pérez, who suggested several mental health diagnostics that could be added to our tool. These were all off-the-shelf proven tools.

We began by adding a module that was designed to measure stress. We called the diagnostic the Organizational Health Survey. In 1999, we were engaged by a large American retailer to use the Organizational Health Survey. When the survey results were available, we passed the data to the Research Services Unit at the Faculty of Medicine at the University of Toronto, to run principle component analysis and correlates so that we could see if there were any linkages between the organizational factors and the stress measures. Their findings were supported by our later research, which is presented in this book.

Developing the Employee Engagement Survey

Around this time, Entec added to its team a number of experts in several areas: Hammo Hammond, an expert in organizational development, who, up to this point, was a key organizational development resource to Volkswagen for well over six years; and Dr. Dalton Kehoe, professor in leadership and communication at York University in Toronto, who brought the book on employee engagement, *First Break All the Rules* by Buckingham and Coffman, to our attention. In this book, Buckingham and Coffman define employee engagement as employees who are committed to and passionate about their work. Others who wrote about employee engagement expanded the definition by adding employees who were emotionally connected to their organization and to their coworkers.

In the late nineties, as part of our research into organizational health, Entec Corporation was examining a flood of statistics that showed the rapid increase in the cost of absenteeism as it related to mental disabilities. For example, in 1998, in a telecommunications company with thirty-four thousand employees, the cost of absenteeism due to mental disabilities had reached 35 percent of the total cost of absences.

We also examined drug use statistics at various companies and noted a significant increase in drug usage for various types of emotional disabilities. At one of our client organizations—a hospital with thirty-five hundred employees—between 1997 and 2001, there was a five-fold increase in the use of drugs such as Welbutrin©, Prozac©, Zoloft©, Paxil©, and other drugs that are typically used for treating depression, anxiety disorder, and burnout. In their May 1999 report, The Global Business and Economic Roundtable on Addiction and Mental Health estimated that the cost of depression to business in the United States was about $60 billion per year.

From our vantage point, there was a gap between the published research on employee engagement and the accelerated growth in emotional disabilities. It became clear that an understanding of the relationship between mental health and employee engagement was needed. Because of the serious nature of the subject matter we were measuring, it became necessary to add another measure of rigor.

We established a Research Advisory Panel on Organizational Health. The purpose of the panel was to provide guidance and feedback on our work and on the research we were pursuing at the conclusion of ever project. The research advisory group comprised four individuals: Dr. Lynn Holeness, Director of Occupational Health at St. Michael's Hospital, Dr. Bruce Rowat, Vice President & Medical Director, Sun Life Insurance, Dr. Baba Vishwanath, Dean of Business, McMaster University, and Dr. Edgardo Pérez, President and CEO, Homewood Health Centre.

Although the team held only one formal meeting, input and feedback was received from individual members of the advisory group on an ongoing basis. The most profound suggestion offered by the panel came out of our first meeting when Dr. Vishwanath said we needed to develop a model for employee engagement that succinctly defined organizational health. Later, Dr. Don Fulgosi, a psychiatrist, joined the team and provided valuable input into the final construct of the well-being portion of the diagnostic.

Entec assembled a team that included experts in organizational development, strategic management, leadership, behavioral psychology, medicine, and psychiatry. This breadth of professions was needed to ensure that employee engagement was considered from a systems perspective. Their task was to develop a comprehensive model of employee engagement.

It took about two years of research and discussion for the team to create the Employee Engagement Model© (depicted below). Four components were thought to influence employee engagement:

- Health
- Leadership
- Mission and Values, and
- Organizational Practices and Processes.

The Individual Health module was divided into two parts because it was apparent that there are factors over which an individual has control, and external elements over which they have little or no control. For example, everyone has the opportunity to control their eating habits, physical fitness, spiritual fitness, and other lifestyle factors, such as smoking and alcohol consumption. However, individuals have no control over a host of external factors such as terrorist attacks, job loss, physical injury, physical and psychological abuse, serious illness, and death of a family member or a friend.

Both positive and negative events that come our way are a part of normal life. External events over which we have no control will influence our well-being in direct proportion to the dramatic nature of the event and our ability to cope. Our ability to cope is determined by the choices we make in the areas of our lives over which we do have control. Our ability to cope with serious, unexpected events will depend on our mental, physical, and spiritual fitness, and on our understanding of our personal mission in life.

Similarly, within the organizational context, there are practices over which an employee has varying degrees of control. Most employees have some level of input regarding practices at the team or departmental level. Typically, they have little to no say on corporate-wide policies and practices regarding governance, mission, and business strategy that are the domain of senior management.

When the model was completed, the team was asked to create an employee survey that seamlessly included questions for both the organizational and the individual health portions of the Employee Engagement Model©. The team decided to consider the questions from the QDT© because it gave them a source of validated questions that could be appropriate for the four organizational modules in the model. In addition, each member of the team was able to draw upon their own area of expertise and add questions that were missing. On the health side, the Research Advisory Group proved to be an excellent source of psychosocial and medical research, and known diagnostic tools in the areas of measuring stress, burnout, and other psychiatric disorders. Chapter 6, "The Work/Health Connection," will explain in greater detail the process that the team followed in arriving at the health questions.

Employee Engagement Model©

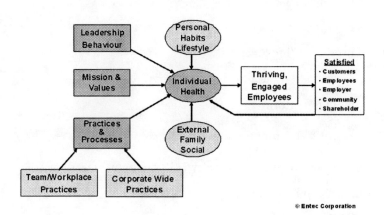

© Entec Corporation

The Employee Engagement Model© made it surprisingly easy to develop the survey, providing logic for the selection and placement of questions. When the survey was completed, the team agreed to name the new survey the Organizational Health & Employee Engagement Survey©.

As time went on, we discovered that not all companies were interested in the comprehensive survey. Organizations had different levels of knowledge and understanding of employee motivation, commitment, and engagement. In fact, a number of companies were not familiar with the term "employee engagement." They were still thinking in terms of "satisfaction." In response to the differing levels of knowledge and need, we de-modularized the survey and created three versions:

Level 1 Employee Engagement Survey (67 questions) measures:

1. Leadership Behavior

2. Mission and Values

3. Team/Departmental Workplace Practices

4. Corporate Wide Practices

This survey takes 6–8 minutes to complete.

Level 2 Employee Engagement Survey (84 questions) measures:

1. Leadership Behavior

2. Mission and Values

3. Team/Departmental Workplace Practices

4. Corporate Wide Practices

5. Mental and Physical Energy (symptoms of burnout)

6. Ability to Perform

This survey takes 8–10 minutes to complete.

Level 3 Employee Engagement Survey (119 questions) measures:

1. Leadership Behavior

2. Mission and Values

3. Team/Departmental Workplace Practices

4. Corporate Wide Practices

5. Mental & Physical Energy (symptoms of burnout)

6. Mood (symptoms of depression)

7. Mental Focus & Self-Worth

8. Control & Lifestyle

9. Ability to Perform

This survey takes 10–12 minutes to complete.

Offering this range of surveys met a wider range of client needs.

The Level 2 survey was developed last. It emerged from research we conducted using client data from the comprehensive Level 3 survey. This research is outlined in Chapter 6, "The Work/Health Connection."

Chapter 2

The Knowledge-based Economy:

Employee Engagement and Mental Health

Organizational health is a continuous process of understanding and fine-tuning the dynamics of the employer/employee relationship to satisfy their mutual desire for corporate and individual prosperity and well-being. This process leads to employee engagement.

—Michael Koscec
January 2000

Employee engagement is a complex subject and use of the term has become ubiquitous. All management consultants refer to employee engagement. Surveying companies have re-branded their employee satisfaction surveys and have renamed them employee engagement surveys. Marcus Buckingham and Curt Coffman in their book, *First Break all the Rules*, brought the term "employee engagement" into the mainstream. They wrote that engaged employees work with passion and feel a profound connection to their company. They drive innovation and move the organization forward.

In their book, Buckingham and Coffman examined a large quantity of data and concluded that the degree of employee engagement could be determined by asking employees twelve questions. All of these questions measured some aspect of the work environment. They were designed to assess various organizational issues as they related to practices and leadership. The answers to these questions made a valuable contribution to furthering our understanding of factors that contribute to employee motivation and employee productivity.

Prior to this time, Entec Corporation had been assisting companies with organizational alignment. Many companies had articulated business objectives, but they were frustrated at their inability to reach those objectives. We discovered that a major contributor to this frustration was the fact that these companies had practices, processes, and policies that were in conflict with each other and with attaining desired objectives. For example, one particular company had sales targets for their sales team. They also set revenue growth targets for each year. Even though sales targets were met, gross revenues and net income were flat year after year.

When we examined the problem at the company, we discovered it was losing as many customers as it was adding each year. Digging further, we discovered that the delivery of the service they were selling was grossly under-funded. The turnover rate of their consultants was an astonishing 50 percent per year. These consultants were highly trained professionals. Their bonus and performance compensation was highly skewed in favor of the company. As their work with clients increased, the company was earning a disproportionately greater share of revenues. It was a great business model on paper, but it didn't work in reality. The turnover rate of clients was remarkably close to the turnover rate of consultants. It was clear that there was significant misalignment between corporate goals and internal systems and practices to achieve those goals.

After this experience, Entec developed an employee survey to measure organizational alignment. It contained many questions that one would expect in an employee survey, but our analysis focused on looking for organizational alignment and misalignment.

From Industry to Information

After several years, we were struck by the compelling data and information that emerged on the side of employee wellness. One of the first books that changed the course of our thinking about employee motivation, morale, and productivity was *Mindsets* by Bill Wilkerson and Dr. Edgardo Pérez. Their book brought to light the high cost of mental disabilities to business and to the economy. They noted that in 1993, on a worldwide basis, expenditures on information technology exceeded expenditures on oil by businesses. In addition, Thomas Stewart, a member of the board of *Fortune Magazine*, in 1991, noted that capital spending in America on machinery and equipment and other technologies for manufacturing was $107 billion, and for information technology, it was $112 billion. This marked the first time in history where spending on information technology outstripped spending on production technology typically associated with jobs in manufacturing.

The significance of these two occurrences is that they mark the crossover from the industrial age into the information age and the knowledge-based economy. Bill Wilkerson coined the notion that today, we have an economy that is based on mental performance, or a "brain economy." This economy is now well rooted, where the value of employees' thought content in generating profits is much greater than the value of the physical output typically associated with manufacturing. In other words, the mind is contributing a much larger share to our standard of living and to the production of goods and services than the physical body. McKinsey and Company estimates that 85 percent of new jobs require cerebral rather than manual skills.

This migration into the higher content of the mind is happening in all sectors, including those that are associated with high-cost capital equipment and labor, such as steel and automobile manufacturing. To illustrate the point in the automobile sector, the Roundtable states that the dollar sales value of the thought content built into cars in the form of design, efficiency, engine management systems, and safety features, exceeds the dollar value of the steel in cars. There are more microchips than spark plugs.

To illustrate this point again, the Roundtable offers an example in the steel industry. Today, changes in technology and the use of computers have reduced the time by half to produce a ton of steel. These are direct outcomes of the greater amount of thought content in the production process.

A more dramatic example of the large proportion of thought content in a product is our desktop computer. Five years ago, I paid $5,000 for my laptop. I had it loaded so that I wouldn't have to replace it for a couple of years: 512 megabytes of RAM, a forty-gigabyte hard drive, and so on. I ordered the newly released Microsoft XP Professional operating system and tons of software that I needed for my work. Today, the same computer with the same capacity and the same software costs approximately $1,600. The price reduction is primarily in the hardware. The thought content in a computer—the software—exceeds the cost of the hardware.

Not surprisingly, this shift from physical labor to mental labor has been accompanied by a shift in the types of absences that are occurring in the workplace. In many organizations today, 40 to 50 percent of the costs of absences from work are directly related to mental disabilities, and this number rises significantly if you add the physical disabilities (such as heart disease) that are associated with stress, burnout, and depression.

Dr. Neva Hilliard, a former director of the British Columbia Workers' Compensation Board, noted that today, most injured workers are absent from work longer, and the cost of caring for them has risen dramatically. Reported Workers' Compensation Board claims in British Columbia were actually down 35 percent over a five-year period, while the cost of claims soared 264 percent. She cited workplace stress as a rising contributor to disability as we enter the era of the white-collar knowledge worker. Workers who were off on stress leave were absent from work on an average of three months. This was a longer period of time than an employee who had undergone heart bypass surgery.

According to the U.S. Health Care Financing Administration, companies were projecting to spend more than 60 percent of their after-tax profits by the year 2000 on the provision of medical care for their employees due to stress-related illnesses. In 2006, reports abound about the high cost of healthcare. A good example is General Motors. In 2004, GM spent $5.6 billion (U.S.) on healthcare and is considered the largest private healthcare provider in the United States. In Canada, Manulife Financial was the largest supplier to General Motors, larger than any of its parts suppliers.

According to a large psychiatric hospital, the workplace is the source of 30 percent of people seeking treatment for depression. When treatment is completed and the individual is deemed healthy enough to return to work, they return to the same environment that created the illness in the first place. This situation is no different than treating a person for a breathing problem created by toxic fumes in a poorly ventilated paint shop, and then returning the individual to the same paint shop to continue to work. The negative outcome of recurring illnesses is predictable. Similarly, non-manufacturing workplaces may be "toxic," and one can expect comparable outcomes when recovered persons return to those toxic workplaces.

The American Institute of Stress estimates that job stress now costs the U.S. economy about $300 billion per year. This is a composite figure representing lost production, medical expenses, workers' compensation, and disability costs.

Stress claims are the fastest-rising category of workers' compensation payouts. The incidence of people suffering from severe anxiety is rapidly increasing due to the higher levels of stress in modern society. About twenty-five million American adults currently suffer from this condition, and it is projected that about sixty-five million Americans (nearly 25 percent of the population) will suffer from an anxiety disorder or depression at some point in their life.

In Canada, depression currently represents 14 percent of all disabilities, greater than the world average. In 1997, in a study by the Homewood Centre for Organizational Health, Canadian workers cited mental anxiety and stress as reasons for growing levels of absenteeism more often than physical illness.

Forty percent of Canadians identify worry and anxiety as their principal de-motivators at work. Twenty percent of individual trips to the family doctor are prompted by stress, depression, and anxiety disorder. *The Global Burden of Disease and Injury Study*, conducted by the Harvard School of Public Health and the World Health Organization, found that in the developing regions, where four-fifths of the planet's people live, non-communicable diseases, such as depression and heart disease, are fast replacing traditional enemies, such as infectious diseases and malnutrition, as the leading causes of disability. By the year 2020, non-communicable diseases are expected to account for seven out of every ten deaths in developing regions, compared with less than half today, and heart disease and depression will be the two leading causes of disease burden.

In their landmark *Work-Life Conflict Study*, Linda Duxbury and Chris Higgins discovered that over a ten-year period between 1991 and 2001, the number of employees who worked between forty and forty-four hours increased by 4 percent, from 30 percent to 34 percent; those working from forty-five to forty-nine hours increased by 4 percent, from 11 percent to 15 percent; and those working fifty-plus hours increased by 16 percent, from 11 percent to 27 percent. So it's no wonder that 40 to 50 percent of the costs of absences from work are directly related to mental disabilities, and this number rises significantly if you add the physical disabilities (such as heart disease) that are associated with stress, burnout, and depression. In 2006, Warren Shepell, an employee assistance provider (EAP), reported that disability claims for mental disorders had risen by 40 percent in the past decade.

The statistical evidence describing the seriousness of the mental health problem as it relates to the workplace is overwhelming. These statistics convinced us that we needed a new definition of employee engagement that considered the employee from a much broader perspective.

A Holistic View of Employee Engagement

Based on the changes in the nature of work, we developed a new definition of employee engagement: Employees who are engaged are mentally and physically healthy, and they are passionate about their work, and are emotionally committed to their organization.

We also felt that there was a need to define leadership within the context of employee engagement: Leadership is the ability of managers at all levels of the organization to create working conditions that will allow their employees to function at peak performance, a working environment where all employees can thrive and be fully engaged.

If we consider these two definitions, it becomes evident that there are two sides to employee engagement: the emotional, psychological, and physical health of the individual employee, and the working conditions in the organization created by managers at all levels of the organization.

The flip side of employee engagement is employee disengagement. These are employees who are at work physically, but who are not there mentally. Their minds are not on their work for various reasons: they are in emotional or physical distress (i.e., genuinely suffering from an illness); they're bored; their mind is on last night's date; they're thinking about a sick family member; they're planning the tail-gate party for next weekend's football game; they're experiencing financial problems because their spouse was laid off; they're anxious about their own job because they're not able to complete all of their work on time; they're afraid of their immediate supervisor or the changes that are taking place at work; and they're fearful of losing their jobs.

There are as many reasons as there are disengaged employees for not being present in their jobs. The term "presenteeism" preceded the term disengagement. Several years before the publication of *First Break All the Rules*, my clients and colleagues used to talk about the problem of presenteeism as a problem that was costing companies more money in lost productivity than absenteeism. In the November 2004 issue of the *Harvard Business Review*, a study on the cost of presenteeism to a U.S. business was published. The results were summarized in an effective graph, which is shown below. The research showed that the direct medical cost of sick employees to their organization was 24 percent, and the indirect cost was 76 percent. The indirect cost included 6 percent for absenteeism, 6 percent for short-term disability, 1 percent for long-term disability, and a whopping 63 percent for presenteeism. In other words, fully 63 percent of the cost of lost productivity to this U.S. business could be attributed to employees who were at work, but whose minds were not on their jobs.

In this particular study, the biggest contributors to presenteeism, in terms of cost to the business, in descending order were:

- Allergies and sinus problems
- Arthritis
- Chronic lower-back pain
- Depression

- Dermatitis or other skin conditions
- Flu

This represents one cost aspect of presenteeism—the medical perspective. It does not capture the non-medical perspective, where an employee is at work but not focused on work. The employee is thinking or daydreaming about: weekend plans, the big game, a date, what to prepare for dinner, or just worrying about their job, children, spouse, or sick friend. All of these are real and, if they occur periodically, are not bad. But if this is a continuous state of mind for an employee, the cost to productivity is significant.

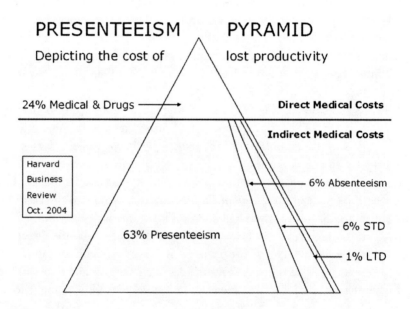

Interpreting Employee Engagement

Entec's Level 3 Employee Engagement Survey© measures organizational practices, leadership behaviors, mission and values, and employee well-being. The organizational factors create the specific working conditions that influence an employee to be motivated, and to be emotionally committed to their work and to their company. Since every employee is unique, psychologically, emotionally, socially, and circumstantially, each employee brings a unique set of needs, wants, and expectations to the workplace. Each employee will respond differently to the same working conditions and work-related stressors.

For example, every employee has a different level of self-motivation. One employee may require verbal recognition once a year for a job well done, while another employee may require recognition once a week. Each of these employees will score the question regarding recognition differently, even though they may have the same supervisor and are treated the same way.

If 36 percent of employees scored in the disengaged category, it means that the organizational practices and leadership behaviors are not meeting their needs and motivating them to be fully engaged. It does not mean they are "slugs." It means they need more from their organization to lift their level of performance. Disengaged employees can become engaged employees under the right conditions.

It is important to convey to all employees that the "disengagement" score is not a negative reflection of their own desire to do a good job. It means that the organization has failed to create the right conditions for encouraging employees to reach their full potential.

But employees also need to understand that employee engagement is a partnership between themselves and the company. The responsibility for employee engagement does not rest solely on the shoulders of the organization. It is not one or the other—it is both. Employees must meet the needs of the company. They also have to take responsibility for their own self-fulfillment within the context of their work. Employee engagement is a partnership between local people managers, senior management and the employees, where everyone works together to achieve the business objectives of the company and the personal aspirations of the employees. However, this is easier to achieve if senior management and local people managers first create the conditions for this to happen.

Changes in Employee Engagement

Before we show differences in employee engagement between good and poor organizations and between various job levels, we need to say a word about our employee engagement categories. Working with our clients, we soon discovered that the three engagement categories used by Gallup (engaged, disengaged, and actively disengaged, where only employees who assigned a score of five on a five-point scale were deemed engaged) did not adequately reflect the sensitivity between good organizations and poor organizations. Their categories did not reflect the realities of our clients. For example, we worked with clients who were among the top organizations in their sectors in terms of financial performance, market share, and other measures, such as rates of employee turnover, absentee-ism, and medical costs. Using Gallup's standard of five (for a completed employee survey) to represent an engaged employee, there would have been very few engaged employees, far fewer than Gallup's 26 percent in the United States.

This could partially be explained by the differences in our surveys. Entec's surveys contain more organizational questions, and they contain questions measuring employee well-being. To reflect the reality of our clients, and to compensate for the difference in our survey, we added a fourth category: actively engaged. The following table depicts the scoring for the four employee engagement categories.

EMPLOYEE ENGAGEMENT CATEGORIES		
Category*	**Scores**	**Description**
ACTIVELY ENGAGED	4.5-5.0	Employees are emotionally committed to their work and to their organization all of the time. They are highly self-motivated and really want to make a difference. They always exceed performance expectations. Their rate of absenteeism is well below industry averages.
ENGAGED	4.0-4.49	Employees are emotionally committed to their work and to their organization most of the time. They always meet and sometimes exceed performance expectations. Their rate of absenteeism is somewhat below industry averages.
DISENGAGED	3.0-3.9	Employees in this category show up to work regularly, but do just enough to get by. This group has been labeled by the term "presenteeism." They are at work physically, but their minds are elsewhere. They like to take advantage of their allotted absent days.
ACTIVELY DISENGAGED	1.0-2.9	Employees are disinterested in their work and in their organization, and they actively work to undermine their fellow employees and their company by spreading malicious rumors, gossip, complaining, and poor performance. They take full advantage of their allotted absent days, and in some cases exceed them.

* These categories are based on a basket of companies, including pharmaceutical, manufacturing, distribution, high-tech, service, retail, healthcare, education, gas, and electric utilities.

The following three pie charts depict changes in employee engagement. They show the results of the best organizations in our database, those in the middle, and the worst. This database comprises twenty-eight organizations, representing approximately forty-five thousand employees.

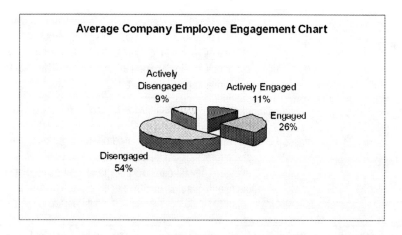

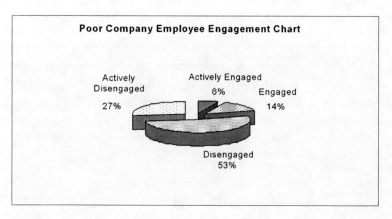

The bookends of employee engagement: In good companies, there are always actively disengaged employees. In poor companies, there are always actively engaged employees.

The first obvious insight that one can glean from these charts is that in very good organizations there are actively disengaged employees. The actively disengaged employees represent 7 percent of the total employee population. At the other end of the spectrum, in very poor organizations, the actively engaged employees represent 6 percent of the total employee population. The two types of organizations are almost mirror images of each other. The following table compares the levels of employee engagement:

Level of Employee Engagement	Good Company	Average Company	Poor Company
Actively engaged	14	11	6
Engaged	37	26	14
Disengaged	42	54	53
Actively disengaged	7	9	27

Discovering that they had a hardcore group of actively disengaged employees was one of the biggest lessons for our clients' human resources departments. When we discussed these findings with many of our clients, we discovered that a small group of employees take up the largest share of an HR department's time. There were no discernable differences between non-union organizations and organizations with unions. The HR departments learned that some employees in their organizations were psychologically wired to be negative, uncooperative, and generally dissatisfied with their lot in life. Within the context of organizations, there is very little that can be done to "change" these employees. It's just who they are. We specifically say "within the context of organizations" because some will argue that everyone can change. They will refer to hardened criminals who have transformed their lives. This may be true; however, there are limits to an organization's ability and responsibility regarding people's personal lives. These responsibilities are distinct from the "rights to accommodate" for mental or physical disabilities.

As noted above, even the very worst organizations have actively engaged employees. That's simply how these employees are wired. They are naturally inclined to be self-starters, highly motivated, and very positive people. Regardless of how bad their organizations are, these employees will excel. They may not stay in the organization long term, but while they are there, they will work hard and they will succeed.

As we move from the good company to the average company, we can see that the widest "swing" takes place in the middle, between engaged employees and disengaged employees. There is an 11 percent decrease in the number of engaged employees in the average company and a corresponding 12 percent increase in the number of disengaged employees. This is especially noteworthy because a significant shift in employee engagement will have an impact on productivity and corporate performance.

Comparing the number of employees that are actively disengaged in a very good organization to an average organization shows that there is a small increase of 2 percent in the average organization. The relative stability in numbers of actively disengaged employees as the quality of the work environment deteriorates, confirms the notion that there is a hardcore group of dissident individuals in our workplaces. In our values, Entec Corporation states: "We believe that the majority of employees seek purpose and meaning in their work, and that they want to make a valuable contribution to their organizations." The lack of movement into the actively disengaged category in response to deteriorating working conditions appears to confirm our belief.

However, the doubling in numbers of the actively disengaged employees in the worst companies shows that there are limits to employee goodwill. Poorly run organizations will create dissident employees. They create their own people problems.

We should also point out another interesting phenomenon. In average organizations, the actively engaged employees have been reduced by almost half when compared to the very good organizations. However, the number of engaged employees remains almost the same. It appears that there is a shift of approximately the same number of employees from actively engaged to engaged, and from engaged to disengaged. The number of actively engaged employees holds fast as one moves from an average organization to a poor organization.

Employee Engagement Between Job Levels

The next set of charts shows the differences in employee engagement between job levels. These charts were chosen from one of the good organizations in our database.

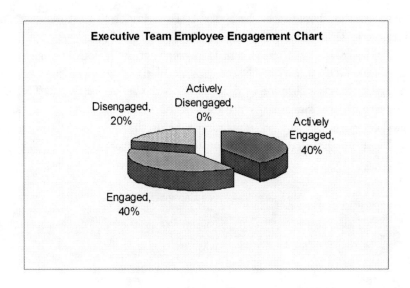

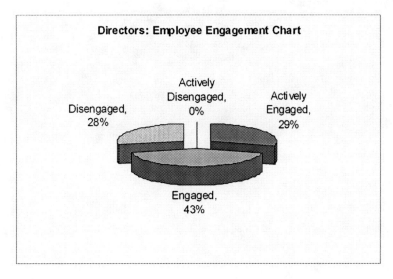

The first chart shows the results of the executive team. Actively engaged and engaged together accounted for 80 percent of the team. To our surprise, even at the senior management level there was 20 percent disengagement. Looking at this data together with our knowledge of the senior executives, we could come up with two explanations. The first was that some senior executives were very close to retirement. There may have been some slowing down as a result of a long life of work and the inevitable downshifting and coasting to the finish line. The other observation we made was that there was evidence of the "Peter Principle" at work where, over the years, long-tenured individuals were promoted beyond their level of comfort.

At the directors' level there was an 11 percent decline in actively engaged. The engaged remained approximately the same, and the disengaged increased by a percentage that was very close to the decline in the actively engaged. This trend continues with the managers, with another 12 percent decline in the actively engaged category, the engaged remaining roughly the same, and the disengaged increasing a little less than the decline in the actively engaged.

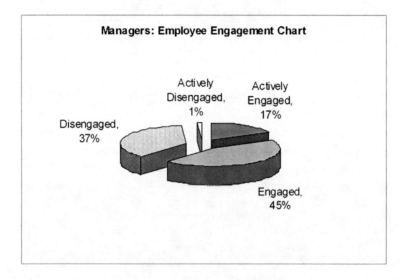

Managers: Employee Engagement Chart

Actively Disengaged, 1%

Actively Engaged, 17%

Disengaged, 37%

Engaged, 45%

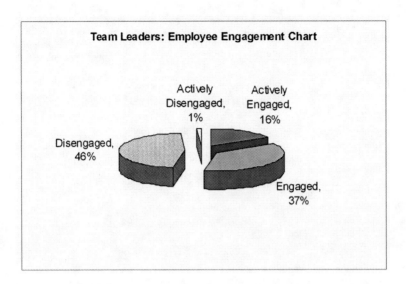

At the team leader level, the actively engaged segment holds firm. However, the number of disengaged team leaders increases by approximately the same amount as the decrease in the engaged numbers. The following table compares these percentages:

Level of Engagement	Executive Team	Directors	Managers	Team Leaders
Actively engaged	40	29	17	16
Engaged	40	43	45	37
Disengaged	20	28	37	46
Actively disengaged	0	0	1	1

The slide in actively engaged and engaged employees continues with the professional, technical, and admin/clerical category of employees. This is especially significant because the frontline workers have the greatest affect on the level of quality of products and services. Their level of engagement determines the level of customer satisfaction and customer loyalty.

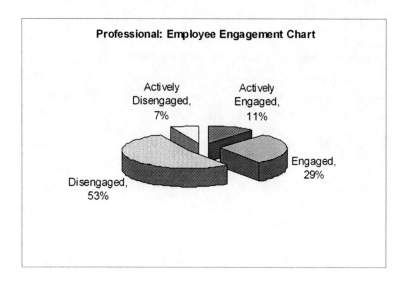

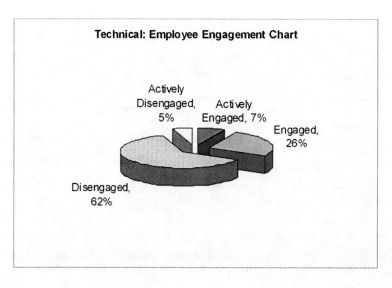

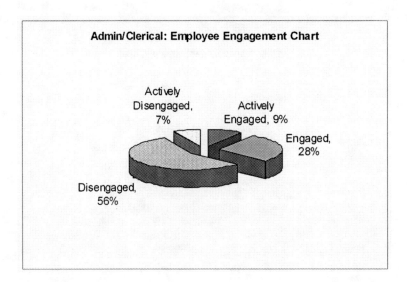

Admin/Clerical: Employee Engagement Chart

For comparison purposes, the following charts are from a weaker organization that is currently working very hard to improve. The charts portray a similar picture of decreasing employee engagement from the top levels of the organization down to the frontline workers.

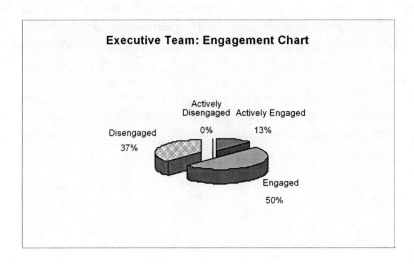

Executive Team: Engagement Chart

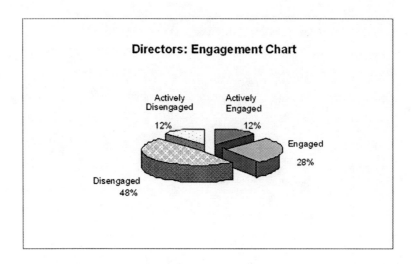

In this particular example, there are a couple of unions. There is a striking difference in the number of actively disengaged employees between the non-union and union employees. The large number of disengaged employees among the union ranks was no accident. They were not disengaged because they were in a union. They were disengaged because of untrained supervisors. The leadership scores of the frontline supervisors who were managing the union staff were very low. As it turned out, none of the frontline supervisors had any leadership training. Their approach to management was to copy the leadership style of their predecessors, who never received management or leadership training. These untrained supervisors were responsible for their local practices. It was not surprising to see that the departmental scores were also very low. The organization was perpetuating a culture of incompetence with the frontline supervisors. They, in turn, were perpetuating inefficient and wasteful workplace practices that frustrated and de-motivated the frontline employees.

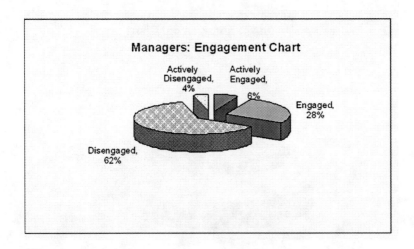

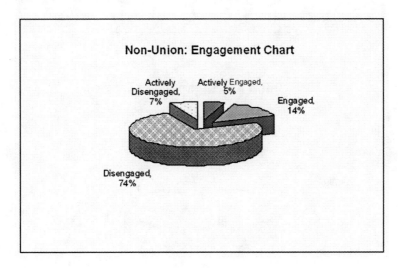

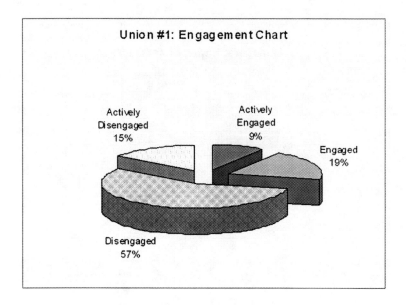

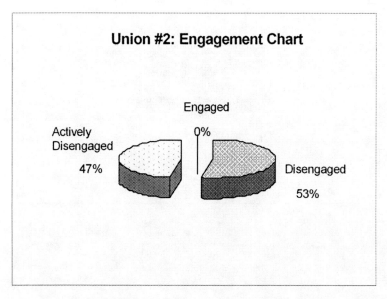

The charts from both the stronger and weaker organizations illustrate the point made earlier about the natural tendency of some employees to be self-motivated, while others have a stronger tendency to be dissident. Based on our research, we can conclude that it is not an accident that there are more actively engaged employees at the top of the organization. However, having said that, and discounting the 6 to 8 percent of employees who are hardcore actively disengaged, one should not expect to have full engagement with the remaining employees. There is nothing wrong with having employees who just want to show up to work, do their job, and go home. Every organization needs "basic" workers. It's the manager's job to make sure that the right working conditions are created. Everyone cannot and should not be expected to be a superstar. Our research indicates that even in the best companies, a level of 40 percent disengaged employees is about the best they can achieve. There seems to be a natural point of resistance somewhere between 30 and 40 percent, beyond which employee disengagement cannot break out.

A Vision for Work/Life Balance

The issue of work/life balance cannot be ignored in any discussion of employee engagement. This is a complex issue because it is very subjective. The first thing employers must realize is that work/life balance is different for everyone. For example, a young married mother with two children is facing different pressures and challenges than a middle-aged employee with teenagers in high school or university and aging parents. Work/life balance for a young single employee would look completely different again.

These differences are further amplified by the nature of work. I know of one couple that works for the same company on the assembly line in the automotive sector. They have two pre-school children. They are able to arrange different shifts for themselves so that one works days while the other works nights. Since finances are tight, this arrangement allows each of them to take turns being with their children while the other spouse is working.

The work dynamics in a financial institution are completely different than in manufacturing, and require a different approach. The complexities of work/life balance are enormous when one begins to think of the varied customer needs and expectations, the many products and services that are being offered, and the many processes that are required to provide the products and services. Add to this mix, company profitability targets, shareholder expectations, and the varied needs of employees, and it becomes apparent why employers don't appear to be moving very quickly in dealing with work/life issues. It is extremely complex!

The worst possible action a well-intentioned employer can take is to develop a set of policies and programs that attempts to fit the disparate needs of all employees, with a one-size-fits-all approach.

What is an employer to do? I believe the first step is for the employer to stand back and ask two fundamental questions:

- What are my objectives around work/life balance, and what are the outcomes I want to see, both for my employees and for the company?
- Do I fully understand the implications of work/life options on the operational needs of the company?"

The first question should be answered from a values perspective. It should be viewed as a philosophical question that demands a philosophical answer, rather than a bottom-line business response.

It seems to me that we need to start by defining work/life balance. I'll take a page from my own experience. In our company, we view work/life balance and organizational health from a similar perspective, in that they are not static states, but rather, dynamic states. Both work/life balance and organizational health are continuous processes of understanding and fine-tuning the dynamics of the employee/employer relationship to satisfy their mutual interests for individual and corporate well-being and prosperity.

This definition provides the clues to a statement of philosophy that a company could develop for itself: "To satisfy our mutual interests for individual and corporate well-being and prosperity." The strength in this definition is that work/life balance is not viewed negatively where the interests of the employer and employee are at odds. They are not pulling in opposite directions. In fact, the employee's and employer's interests are the same: health and prosperity for both the company and the individual employee.

The question that immediately comes to mind is: "What does that look like?" Again, the answer is not simple. It will be different for every organization. However, if an organization begins with a clear "vision" statement for work/life balance, this statement becomes the basis for the development of the guiding principles that will influence decisions and leadership behaviors. This approach allows for maximum flexibility. It enables local people leaders to decide jointly with their staff the most appropriate action under differing circumstances. Local decisions are made to reflect the complexities of the organization, and they are made within the framework of mutual benefit for the employee and the employer.

There is an important issue that I need to touch upon briefly. A flexible work/life balance policy will only work in an environment where communication, education, workplace practices, and leadership behaviors work together to promote employee engagement. This type of workplace gives managers and employees the clear knowledge of their contribution to the prosperity of their company and to themselves. A team leader or department manager will be limited in arriving at optimum solutions to work/life balance issues without the commitment to this symbiotic relationship that binds the employee and the employer. If an employee wishes to take unfair advantage of a work/life balance policy that is based on a value of satisfying mutual interests, the employee will be less likely to do so if he or she knows exactly the implications of their actions.

To conclude, my suggestion would be to develop a simple vision for work/life balance. With this vision as a cornerstone, a company can develop several guiding principles that managers can use to make decisions locally. They must make sure that all the pieces are in place, where employees and managers understand their contribution to the enterprise, and what that means to their personal well-being and prosperity. As an example from our company, we used our own definition of organizational health as our vision for work/life balance:

> *"To satisfy the mutual interests of the individual employee and Entec Corporate in achieving personal and corporate well-being and prosperity."*

Achieving a successful work/life balance is based on a partnership between Entec Corporation and its employees. There are two sets of guiding principles: one for the company and another for the employees.

Company Guiding Principles

1. Entec's work practices will not place an employee's physical and mental health at risk

2. Entec's work practices will not place undue pressure on the employee's personal life outside the office

3. Entec will accommodate to the best of its abilities an employee's personal needs as they arise

4. Entec will sustain a positive and supportive workplace environment where all employees feel nurtured and respected

5. Entec will encourage learning and personal development

Employee Guiding Principles

1. I will take responsibility for making sure that the needs of my clients are met satisfactorily during my absences from the office

2. I will ensure that my work habits are not placing my physical and mental health in jeopardy

3. I will ensure that my work habits are not interfering with the well-being of my personal/family life

4. I will ensure that my lifestyle (eating habits, physical exercise, personal relationships, spiritual growth) contribute to my physical and mental energy

5. I understand how I contribute to the company's prosperity, and I will make the best effort to achieve the company's profitability targets

These guiding principles are rooted in an adult-to-adult relationship between the company and the employee. They are based on mutual trust and respect. As such, the company is committed to full disclosure of company information that allows employees to make informed decisions. It is committed to supporting the employee to grow in any of the areas the employee feels they need to develop strengths.

The way each employee translates these guiding principles into action will vary. Implicit in these principles is that each employee enjoys a large degree of control over their own work, and they have a good knowledge of the work of their coworkers. These principles encourage the employee to look at their own life, and to be intentional in defining what is important to them. Quite frequently, work/life balance goes askew not because of company demands, but because the employee has not taken the time to examine his or her own values. The employee has not taken the time to think about his or her own life and what that life should look like. These employees frequently run into problems. Some of them become workaholics and don't realize it until changes in their personal life generate "push back" from those who are close to them.

A work/life policy, such as the one suggested above, challenges the employees to think about their lives. It challenges them to confront their deeply held values and beliefs. There are vast numbers of people who don't do this type of self-examination. These are the people who are most likely to experience work/life issues. They intuitively feel that something is not right, or they make decisions that negatively impact their work and their company.

This approach to work/life balance provides the flexibility that is needed to meet most work/life issues. Employees are empowered to make their own arrangements to make the policy work. For example, principle #1 leaves the arrangements up to the employee. They know what is expected, and so they know what needs to be done in order to ensure their own needs and those of the company are met. With this model, employees are treated as adults and full, active participants in the life of their organization. It helps employees to thrive and to be fully engaged.

Chapter 3

Organizational Practices

There is no contest between the company that buys the grudging compliance of its workforce and the company that enjoys the enterprising participation of its employees

—Ricardo Semler, owner of Semco Inc.

Perhaps the greatest determinant for disengaging employees is practices and processes that impede work rather than facilitate it. Let's consider the example of a racing team again. A number of conditions are essential for the pit crew to carry out its tasks at peak performance during a race, when the car comes in for a pit stop. First, there must be absolute clarity as to what needs to be done on the car. Communication is open and clear. It starts with the team leader and the driver getting a complete description of the car's performance and a detailed assessment of what needs to be done. This information is communicated to the pit crew, who stand ready as one body to complete the task. They have the information, the skills, and the tools. When the car comes in, the tires are changed and the car is simultaneously refueled. Seven seconds later, the car is out of the pit and back on the track. The pit crew debriefs on its performance and assesses what went well and what could have been improved upon.

This scenario is similar in a high performance company. Everyone has the information, tools, training, skills, and feedback systems to complete their work to the prescribed standards. An environment is created where creativity is encouraged. Employees are challenged to question current practices and to find better and more effective ways to improve product or service quality, and to reduce costs. The performance measure-and-reward system is designed to support and encourage these desired behaviors. Performance feedback is provided not just once a year, which can be stressful and threatening, but regularly throughout the year. In some organizations, this process doesn't need to exist. The individual always knows where they stand. If there are deficiencies, there is a training plan to support improvement. Even if there are no deficiencies, there are training plans in place to allow the individual to grow. Learning is encouraged by the organization, either through company-sponsored programs or other external consultants, colleges, or university courses.

In Chapter 1, we identified practices that take place at the team or department level, where local people managers have a reasonable amount of control over their work. This is distinct from the practices that have a broader corporate-wide reach that impact most employees in the same way. These later practices and policies come directly under the purview of senior management. Our survey questions are grouped together along these lines of responsibility to show who in the organization has the authority for the follow-up implementation. This chapter will be divided into two parts. The first part will discuss departmental practices, and the second will discuss practices at the corporate level.

Practices: My Department

Departmental practices deal with the work dynamics of a team or department. They define the immediate working conditions of an employee.

- Do employees have the right resources and tools to do their work effectively?
- Do they get personal satisfaction from their work?
- Is their workload reasonable?
- Is there an open flow of communication among coworkers and managers?
- Is there a high level of cooperation among coworkers?
- Do employees celebrate when significant milestones are met or projects completed?

All of these practices have a direct impact on the level of employee engagement, performance, and general well-being.

These are work practices over which local people managers have a fair amount of control. In addition, within the context of a team or a department, employees have a greater ability to contribute to the creation of their work experience. This is a critical point. Entec runs action-planning workshops for clients who want help in creating the follow-up implementation plans that result from the survey.

During every workshop we have run over the past ten years, we reach a point where a participant will raise their hand and say that human resources or senior management are responsible for a particular departmental practice. This comment opens the door to my "victim" speech, in which I challenge all frontline workers to take control of their destiny at work. I tell them that they are not victims. If they don't like something, they must take appropriate steps to initiate change. They need to raise the issue with their coworkers and with their supervisor. The issue is not whether they succeed or fail. The issue is that they must make a sincere attempt to change what they feel needs changing.

This gets us into a discussion of fear: the fear of being embarrassed; the fear of being fired; the fear of rocking the boat or upsetting the status quo; the fear of making enemies. Fear is a powerful motivator for doing the right thing or doing the wrong thing. Nevertheless, conquering our fears is a key to living a liberated and fully satisfying life. To meet this challenge, Entec designed a program called "Conquering Your Fears in the Workplace."

I conclude my "victim" speech by saying that employees, local people managers, and senior managers must realize that creating a healthy working environment, where employees thrive and are fully engaged, is a partnership between management and staff. Everyone must work together to achieve the business objectives of the corporation and the personal aspirations of the employees. However, I acknowledge that it's easier if senior managers first create the conditions for this to happen.

Our research shows that these departmental dynamics have the most significant impact on employee engagement. We discovered that although all four organizational modules load strongly to each other, the "My Department" module had the highest statistical loading with the other three modules (Leadership, Corporate Practices, and Mission and Values). The other three modules did not link to one another as strongly as they did to My Department. Of particular surprise to us was that Mission and Values had a stronger loading with My Department than My Manager (leadership behavior).

Tier One Statements: My Department

As we drilled deeper into the data, within the context of departmental practices, we discovered that the following statements were the most important in defining an employee's experience of and contribution to employee engagement. We refer to these as tier one statements:

- Information is shared openly by my colleagues in my department
- I have easy access to the resources, people, and information I need to do my job
- Workload is distributed equally throughout our department
- Everyone pulls his or her own weight equally
- When faced with a crisis, employees in my department pull together to find solutions
- My coworkers work well together
- There is little to no office gossip in our department
- We regularly celebrate our successes

A quick scan of these statements shows a unifying theme. All employees are united by a shared set of values that are based on fairness and justice. They are working together as adults. They respect and trust each other. They look after each other and support each other. And they celebrate together. In other words, they are working together in community. For millennia, communities have been the most powerful vehicles for creating human cooperation. Traditionally, they were essential for survival.

We have always found our source of identity, meaning, and larger purpose in communities. This is where we spend the largest part of our time. Today, our community is the workplace. Many people derive their identity from their work. Graham Tucker used the term "cooperative community" to describe the most powerful type of employee relationships that deliver the highest levels of performance. This distinguishes a healthy community from an unhealthy community, where cooperation, fairness, and justice are absent.

Open Sharing of Information

The first two tier-one statements address access to information and information sharing.

- Information is shared openly by my colleagues in my department
- I have easy access to the resources, people, and information I need to do my job

The nature and quality of information sharing is a more precise measure of communication than simple statements on communication. Information sharing provides a precise insight into the fundamental reason for communication within a workplace context (i.e., the exchange of information). In our work with clients, poor communication between employee and supervisor, among coworkers, and between senior managers and lower level employees is one of the top issues that is raised over and over again. The issue of poor communication has been repeated in the business press and in business books for as long as I can remember. Our research shows that the nature of information sharing is the first contributor to employee engagement.

Open communication is the free flow of information up and down the organization, as well as sideways. It is a flow of information that breaks through all corporate barriers. An excellent example is at Semco, the company that was transformed by Ricardo Semler. No information is hidden in the company. Each month, the company's financial statements are displayed for all to see. Everyone shares in the success or failure of the company. It is in everyone's interest to share information. It frees up time to spend on productivity enhancement, problem solving, and creativity, rather than on company politics and power struggles that are frequently ignited by information withholding and secrecy. Semler wrote a book called *Maverick*, in which he detailed how he changed the company. I would recommend the book to everyone. Suffice it to say that Semco's profits went up by 500 percent in the last ten years. The company employs three thousand workers, yet their turnover is only 1 percent per year.

There are a number of benefits to information sharing within a corporate setting. If one views communication from a "corporate level" perspective, the first level would be corporate communication emanating from the chairman's or president's office, and dealing with corporate change, financial performance, strategic, and policy matters. The way in which this information flows down the organization will impact employee engagement and performance.

For example, is the rumor mill rife with information about corporate changes, or do the senior managers inform employees on corporate matters quickly and directly? Do senior managers view their employees as adults who can responsibly handle information, or as children who need to be protected from sensitive information, and who are, therefore, provided with a carefully rinsed version of facts? As time passes in this latter case, the information will slowly flow to the surface and continue to fuel the rumor mill, increase uncertainty, and lower productivity.

Depending on the size of the organization and the number of levels in the corporate structure, the next level of communication could be at the divisional level, department level, or team level. The same principles apply at the divisional level as at the corporate level. However, what begins to change as one moves down the corporate structure, is the medium used for communication. For example, in a large multinational corporation, the CEO cannot communicate with all the staff simultaneously in person. With today's technology, video transmission is available and commonly used to transmit important high-level messages from a CEO or president.

At the department or team level, where face-to-face meetings are physically possible and are a common practice, are these meetings restricted to disseminating information from the upper level to the lower levels, or are they also used to receive important feedback? At this level, an essential element of employee engagement is the flow of information from the frontline employees back up to the higher levels of the organization. How freely is this information allowed to flow? Is there a filtering that takes place for personal or political reasons?

Near the end of the old Ontario Hydro, a consultant's report by Carl Andognini, a U.S. nuclear expert, was released on the nuclear division of Ontario Hydro. The report concluded that the reactors were technically sound, but that Ontario Hydro had a management problem. Upon further digging, they discovered that crucial reactor performance information, or operator suggestions and observations, were not being passed up the line. The reason cited was that there was a culture of not rocking the boat, not questioning the authority of superiors, and, therefore, not putting one's career and promotion prospects in jeopardy.

The net result was that the president immediately resigned, and seven of the nineteen reactors were laid up, which cost many billions of dollars in repairs and supplemental coal burning. This is a huge price to pay for poor communication and management practices. Beyond the financial consequences, the incident eroded the credibility of Ontario Hydro, and placed in question the whole nuclear option for generating electricity.

At the lowest level, we must look at communication among colleagues, in teams, and between teams or departments. Many organizations have abandoned the traditional departmental structure and moved to self-managed teams. These teams are formed around processes that are totally focused on providing a service or product to a specific customer group. Communication within this team framework is completely open. There is no withholding of information because it would jeopardize the team's success. The individual succeeds only if the team succeeds.

However, the team organization cannot always remain completely "pure" in its one-customer group focus, and must share resources with other teams. Again, there has to be a willingness on the part of all team members in both camps to keep communication completely open.

There are a number of ways to ensure that this happens. One of the most successful has been the involvement of teams in the business planning process. By jointly participating in creating the business plan, the teams know and agree ahead of time on sharing resources and on timing. Through the business planning process, they have a better appreciation of how they fit into the bigger company picture, as well as how other teams contribute and fit in.

There is excellent software available that pyramids the whole planning process, from the individual task and cost components, through the teams, and consolidated right up to the company level. When this system is accessible by all the teams, as well as the corporate office, they see for themselves the output of their participation in the business planning process and can track their progress as well as that of other teams. This approach to creating and managing the business plans "operationalizes" the communication component so that it is not left to chance. There is no mystery where the company is going and how it is doing. Everyone knows how everyone else is progressing. It's an instantaneous performance-feedback mechanism. During meetings, they can focus on discussion of problems and solutions if milestones are not on target. The information is there for all to see.

Equal Distribution of Workload

Although the My Department section contains statements about workload, the statements that rose to the top were those that addressed equal or fair distribution of workload:

- Workload is distributed equally throughout our department
- Everyone pulls his or her own weight equally

Several years ago, research by Martin Shane and other scientists in Scandinavia showed that there was a direct relationship between fairness and trust in an organization and the ability of an employee to handle stress. Incidents of stress-related disorders, such as mental disabilities, heart disease, cancer, and infections, were anywhere from three to five times higher in a workplace where levels of fairness and trust were low. Our research indicates that fairness in decision making, as it relates to workload, has a direct affect on the levels of employee engagement as well. Our own research, as well as published research, shows the inexorable link between employee engagement and employee health.

Cooperation

The next two tier-one statements address employee cooperation:

- When faced with a crisis, employees in my department pull together to find solutions
- My coworkers work well together

In some respects, this should not be a surprise. Open sharing of information would not be possible without a desire by employees to cooperate with each other. More and more companies are beginning to appreciate the advantages and power of working in teams. A few years ago, we were working with a client in the retail sector whose company periodically ran competitions for their sales staff. The typical approach pitted individual salespeople against each other. But sales did not rise appreciably during the duration of the sales competition. At the same time, employee engagement actually dropped. We shared data with the client that showed cooperation was valued over competition. The next time a sales incentive program was launched, employees were grouped in teams, and teams competed against each other. Sales rose and employee engagement improved.

I experienced a similar situation at Ontario Hydro. When I set up the satellite offices, the property agents had an opportunity to interact with their fellow agents and their supervisor at the start of each day. They were able to discuss and solve problems together. Their motivation and commitment to the work rose now that they were part of a community and they no longer worked in isolation.

A principle of community is that people are stronger together than they are separately. For example, seniors who are a part of a community (i.e., they have friends that they see regularly and are regularly involved in group activities) are far healthier and happier than seniors who are isolated. Community develops in an organization when there is a high level of trust, in which supportive relationships can develop. Cooperation is the outward manifestation of a community environment that is built on trust. In the next chapter on leadership, we will look at the importance of trust in employee engagement.

Gossip

We were not surprised to see that gossip was in the top tier of practices that affected employee engagement:

- There is little to no office gossip in our department

Gossip can be a substantial de-motivator in the workplace. Gossip thrives where communication is poor. In an example cited later in this chapter, where six utilities were being amalgamated, gossip was virtually absent from our client utility. There was a great deal of uncertainty, but through open, honest, and frequent communication, gossip was kept to a minimum. Negativity and other destructive behaviors that are fostered by gossip were absent.

Celebration

We work with many great organizations and have observed that celebrating successes, achievements, and milestones rarely occurs.

- We regularly celebrate our successes

My daughter is a chartered accountant who became a partner with her firm at the tender age of thirty-two. When she started working at the firm—which employed about forty people—shortly after graduating from university, she told me how much she enjoyed working there. When I asked her for specific examples, she told me how caring and personable her immediate supervisor and the four partners were. The whole working environment was very positive.

The event that impressed me the most occurred the day before she had to write her Uniform Financial Institute (UFI) exams. These exams must be written to receive a CA designation, and anyone who is familiar with the UFI exams knows that it frequently takes two or three attempts before a person passes the exams. Late that afternoon, everyone gathered outside her office, and one of the partners came down the hall with a very large cake. On the cake was an inscription: "Congratulations Michelle and Good Luck Tomorrow." Before cutting the cake, one of the partners said a few words. He wanted her to know that they were all rooting for her and celebrating the fact that she was going for her CA designation. It was a huge boost to her self-esteem and overall motivation.

After she had her CA designation and assumed a management position within the firm, my daughter was approached by a headhunter at least once a month and offered higher compensation and more responsibility, but she loved her firm and remained loyal to them. Her commitment paid off. She became a full partner at a young age. There is a fairly high "churn" rate among CA professionals, especially with the newly minted, young, and aspiring chartered accounts. My daughter's firm has one of the lowest turnover rates in the CA community. Celebrating pays off.

Tier Two Statements: My Department

There are six more statements that have high loading factors, but are not as strong as the tier one statements. These practices also figure prominently in defining a positive working experience that contributes to employee engagement. Tier two statements deal with personal fulfillment, cooperation with managers, fairness in work scheduling, and participation in decision making. In order of priority, they are:

- I am able to contribute what I do best every day
- I rarely experience conflicts with the managers I work with
- The scheduling of my work is fair and reasonable
- I feel that I am a success at work
- I have the right tools/equipment to do my job the best I can
- I participate in decisions that impact all aspects of the work of my department

Some of the tier two statements are a further extension of statements in the tier one list. For example, the statement: "I have the right tools/equipment to do my job the best I can," is related to the statement: "I have easy access to the resources, people, and information I need to do my job." In addition, the statement: "The scheduling of my work is fair and reasonable," is associated with "Workload is distributed equally throughout our department," and "Everyone pulls his or her own weight equally."

Despite the close association among some of these statements, they are distinct enough to add valuable insights to the dynamics of the practices at the team and/or department level. There are seven other questions in our engagement survey in the My Department module. However, the fourteen statements of practices listed above have the greatest impact on employee engagement. Of great significance is that all of these practices are under the control of the local people manager. The local people manager has the opportunity to create a healthy working environment by zeroing in and working toward the adoption of these practices in his or her department.

I'd like to raise another important issue at this point. During the post-survey action-planning workshop that we offer our clients, one issue always emerges without fail. These workshops typically involve 50 percent frontline staff and 50 percent managers at all levels of the organization. The frontline staff will always point the finger at senior management. From their perspective, the senior manager is responsible for everything, including local practices.

At this point, I usually call a time-out. I deliver my standard speech, pointing out that frontline staff are not victims. I point out that frontline staff have more power than they realize, and there is nothing to prevent them from talking to their immediate supervisor about ways to improve local practices. I also point out that employees need to take some responsibility, and take an active role in improving their own working experience in partnership with their manager. This cannot be stressed enough. And the local people manager should also take a proactive approach in educating his or her employees to become active participants on this journey.

When I think back to the work environment that I inherited at Ontario Hydro, it had all the makings of a perfect storm—a storm of immense destructive power. Most of the foregoing tier one and tier two practices were absent. No consideration was given to workload. Property agents were given their lists of purchases and that was all. Communication flowed only in one direction: from Head Office to the field. The communication was project-specific. Any communication regarding corporate strategy, issues, or changes was learned from reading the newspapers. The property agents could not do what they did best, and they did not feel successful. Conflicts with their immediate supervisors and head office managers flourished. Their work was not valued because they were frequently directed to violate the standards established by their profession. Gossip abounded and there was never a reason to celebrate. These workplace practices proved to be destructive. They contributed to low morale, low productivity, high costs, alcoholism, strokes, and a heart attack.

Corporate Practices and Policies

Corporate practices and policies are under the full control of senior management. They affect all employees equally. In this section, we will attempt to answer questions about the importance of compensation, incentives, personal development, fair promotions, corporate communication, work/life balance, and rate of change to employee engagement. We will begin by looking at the statements on corporate practices that have the highest loading factors (.700 or higher) in the Corporate module. In order of importance they are:

- Candidates that are best suited for the job receive promotions when job openings become available

- I am provided with ample opportunities for personal development

- My organization provides the right amount of information (for my needs) on its direction and performance

- Our incentive and reward system motivates me to do my best work

- The initiatives offered by our organization are appropriate for my health and wellness needs

The first statement reinforces the importance of fairness and trust. Job promotions that are seen as being fair raise the level of trust among employees, and this, in turn, lifts employee engagement.

The importance of personal development has been documented by many research studies over the last few years as being critical to the retention and motivation of employees.

Open communication in the context of the corporation doesn't rank as high as it did in the context of an employee's department. Nevertheless, it is still near the top, demonstrating the importance of open communication to employees, whether it is at the corporate level or the department level.

Most research in the field of employee engagement has found that salary is not that important, as long as it is within the norms of the market. Our research confirms this. However, it's interesting to note that incentives and rewards rank high on the list. Clearly, employees want to be recognized for doing good work. The incentives and rewards don't have to be monetary; however, some kind of recognition that goes beyond regular work-related feedback, such as a dinner for two, public recognition, time off, tickets to the theater, or to a sporting event, go a long way to garner employee engagement.

Health and wellness, as distinct from health and safety, has evolved over the past five years as a legitimate and important corporate concern. Depending on the sophistication of the organization, health and wellness initiatives can be as basic as providing programs for weight loss, quitting smoking, nutrition, physical fitness, and more sophisticated programs on stress management, and recognition and identification of mental disabilities. Health and wellness diagnostics (such as those offered by Entec Corporation), training, and education are becoming more common. Employee assistance programs (EAP) are becoming more specialized and covering a much wider range of counseling and support services than in the past. All of this has arisen out of the recognition that employee health is directly linked to employee performance.

This is especially critical today, because the lean and mean years of the early 90s have not passed. They have become the norm. Continuous cost cutting is a part of everyday working life. It is the new reality. There are two classes of employable people today: those who have jobs and those who don't have jobs. Those who don't are typically middle-aged to mature managers and senior executives who have lost their jobs and cannot find new ones. Those who have jobs are working longer than ever, and the time-freeing technologies like the Internet, cell phones, and Blackberries have become an umbilical cord to the office. The working day has been extended from eight hours to sixteen, and the working week from five days to seven. Nevertheless, we were surprised to discover that the load factor for our statement on health and wellness surpassed that of the "work/life balance" statement.

The following statements on corporate practices still had relatively high loading factors, but were less important than the top five:

- I am fairly compensated for the work that I do
- My organization is reasonable in allowing me to balance my work with my personal life
- I know how changes at our company will affect me personally
- I can cope with the pace of change at our company
- Our internal work processes and procedures are simple and effective
- Our benefits package is appropriate for my needs

This list provides additional insight into the ranking of corporate practices. It is interesting to see that compensation tops this "less important" list. The operative word in this statement on compensation is the word "fair." Research has shown that there are other factors that are more important than compensation; however, "fair" compensation ranked in the middle of all of the statements we measured in this module.

A great deal has been written about work/life balance over the past few years. For example, a survey released by Ipsos-Reid on May 18, 2006, reported that doctors and nurses are at the top of the list of people who cannot take a vacation. The *Globe & Mail* quoted the head of the Ontario Nurses' Association as saying that if you want a day off, you have to ask someone to fill in as a favor, and if you want two weeks off, well, just forget it.

Several years ago, Entec Corporation conducted an employee engagement survey at one of the large Metro Toronto hospitals. The survey indicated low levels of employee engagement, especially among the frontline healthcare workers. It showed that the low levels of engagement were in large part due to burnout by nurses and other frontline workers. The survey showed that 20 percent of staff were already burnt out or were at high risk of burnout.

To "anchor" the survey with the hospital's own statistical information, we examined the hospital's absenteeism records, and discovered that absenteeism was running at fifteen person days per year. That is in excess of two weeks. So, on the one hand, healthcare workers could not take a "planned" two-week vacation, but on the other hand, they were away from work on an "unplanned" basis for over two weeks each year. Clearly, there was something wrong with this picture.

When we discussed this situation with the president, he said there wasn't much that could be done. He didn't have the budget to hire extra staff. We asked him if he knew the cost of high absenteeism, and he said he did not. Shortly after we completed our survey at the hospital, the annual hospital report card on patient satisfaction was released. This hospital was ranked near the bottom of the list. The employees were disengaged, and the patients were not satisfied with the care they received. It was a lose-lose situation. The results should not have been a surprise. If over 20 percent of frontline staff were burnt out, they were incapable of responding to patient needs. They had neither the physical nor the mental energy to carry out their work.

The next two statements in this list ("I know how changes at our company will affect me personally" and "I can cope with the pace of change at our company") address the quality of an organization's information on change, as well as the employee's capability to cope with the pace of change. Managing change effectively is perhaps one of the most difficult aspects of today's business environment. Never have the external factors influencing business been changing as rapidly as they are now. Technology is constantly redefining and forcing changes in manufacturing, communication and office processes. New markets are being created regularly and old markets cease to exist. The rules of the game frequently change as companies pursue global markets.

How do you create an organization that can respond quickly to market shifts, and yet maintain its integrity and a sense of security for its employees? As we will discuss in the next chapter, one answer lies in organizational values, and the importance of having a strong and central ideology binding the organization together and keeping it on an even keel, regardless of how furious and changeable the winds may be. Along with open communication, values that are lived in the organization are the cornerstones of an employee's ability to cope with change.

Several years ago, we worked with a client whose utility was being merged with five other utilities. The terms of the merger were vague. The general managers of the six utilities that were being merged were put in a room and asked to make it happen. No guidelines were given. They were instructed to do it. This was a very fluid situation that took over one year to conclude.

During this time, my client held town hall meetings with all his staff, (four hundred-plus employees), sometimes as often as once every two weeks, but never less than once a month. Entec was asked to conduct an employee engagement survey because the general manager's attitude was "business as usual." We were surprised at the results. The engagement scores were among the highest we had ever seen. Part of the reason was the fact that the general manager was completely open and honest with his staff. Nothing was hidden. Everyone knew as much or as little as the GM. There was little gossip or speculation. The employees trusted their general manager. It had always been that way in the past, and it continued in the current situation. Despite the insecurities surrounding the amalgamation of six utilities, the employees felt secure and were able to cope remarkably well with the changes that were taking place.

Although ineffective internal work processes and procedures ("Our internal work processes and procedures are simple and effective") may become barriers to productivity gains, engaged employees will find ways to work around poorly conceived processes to get the job done. Poor processes are considered a nuisance, but they appear to have little effect on employee engagement.

We were surprised that the statement on the organization's benefits package ("Our benefits package is appropriate for my needs") ranked as the least important. Talking with employees, we learned that benefits are somewhat taken for granted. Most organizations offer reasonable benefits. Benefits are only a concern for workers who don't have them, either because they are self-employed, or they work for very small organizations that cannot afford benefits.

Chapter 4

Leadership

A new moral principle is emerging, which holds that the only authority deserving one's allegiance is that which is freely and knowingly granted by the led to the leader in response to, and in proportion to, the clearly evident servant stature of the leader.

—Robert Greenleaf

Books on leadership line the stacks of libraries around the world. It is not my intention to duplicate any of the information in these books. The focus of this chapter is to discuss leadership from the perspective of employee engagement. I'd like to emphasize that the discussion is on leadership and on leadership behavior as distinct from management.

The reason for zeroing in on leadership behavior is that behavior defines the nature of the point of contact between the manager and the employee. It's at the point of contact that the employee feels the impact of leadership and the employee response is formed. This response is typically not neutral, but rather, it is positive or negative, motivational or de-motivational. A person's behavior is the outward manifestation of inward values and beliefs. Does a leader share the values of Robert Greenleaf, or is the leader motivated by diametrically opposite values of power, wealth, and self-aggrandizement?

Greenleaf maintains that leadership is a state of being, not doing. It is a choice to serve. Without servanthood, leadership is severely limited.

The Difference Between Leadership and Management

Several years ago, I was asked to speak at a conference on organizational health. I asked a colleague to accompany me to this conference. Another speaker at the conference was Louise, who had a PhD in sociology. She was going to speak on leadership and the difference between leadership and management. My colleague and I were eager to hear what Louise had to say. She said that a leader was the CEO, whose job was to look outside the organization and to provide the strategic direction for the organization. Everyone else was a manager. My colleague and I cringed. We had a problem with her description, because we were evangelizing to our clients that all managers at every level should also be leaders.

According to our research from client data and the research of others, we had come to a couple of conclusions. First, leaders are also managers but not all managers are leaders. Leadership competencies tend to favor behavioral characteristics, while management competencies favor organizational competencies. The organizational competencies may include administration, organization, strategy, planning, finance, technology, security, customer service, sales, and marketing. All of the leadership questions in our survey are based on behavioral characteristics. They are questions addressing performance management, communication, personal characteristics, and ethics and justice.

Detailed Research Findings

When employees come to the Leadership module in our survey, they are asked to rate the statements on leadership as they apply to their immediate supervisor (i.e., the person to whom they report). This approach allows immediate subordinates to generate the leadership scores.

When we prepare our survey report, we prepare two types of leadership reports: a direct report and a cumulative report. The direct report contains only the scores of direct reports, so long as there are sufficient numbers of staff to protect confidentiality. The cumulative report comprises all of the leadership scores for the manager's whole line. For example, if a director has five managers and two assistants "directly" reporting to her, she will get a direct report that is comprised of the scores of the seven direct reports. She will also get a cumulative report that is comprised of the scores of all the staff that make up her complete line.

If a "manager" is the lowest level of supervision and has seven (this could be greater in larger companies) or more employees, the manager receives only a direct report. At the lowest level of supervision, the direct report and cumulative report would be the same, so only one report is prepared.

Some of our earlier surveys had well over thirty questions in the leadership section. After every project we ran two sets of analyses: a principle component analysis and correlates. The principle component analysis helped us to cull those questions that were redundant or that had low load factors. It showed only statements that had greater weighting and importance. The correlation analysis identified statements that had the strongest linkages with the other statements in the survey.

Through this process, we reduced the number of leadership questions to twenty-two. Upon further reflection on the questions that remained, we discovered an interesting pattern. We noticed that the questions fell into two overall dimensions: a functional dimension and a values dimension. Within the functional dimension, questions fell into two categories: job performance, and communication and feedback. Under the values dimension, they fell under two categories: ethics/personal characteristics (ethical behavior) and justice/fairness. This sorting led us to develop a model of leadership that we named the Four Dimensions of Leadership.

Admittedly, some questions could be placed in one of two categories. For example, the statement, "Is fair when reviewing my job performance," could be placed in either the justice/fairness category or under the job performance category. We chose to place it under justice/fairness, because our research and the research of others has shown that justice/fairness issues have a much broader impact on employee engagement and performance than issues that are strictly limited to performance.

FOUR DIMENSIONS OF LEADERSHIP

FUNCTIONAL DIMENSION

Job Performance

- Gives me the latitude that I need to do my job to the best of my abilities
- Ensures that I have the right skills and knowledge to do my job to the best of my abilities
- Supports me to develop myself for future advancement opportunities
- Provides me with clear performance expectations
- Reviews my job performance at least once a year

Communication / Feedback

- Listens with an open mind
- Recognizes or praises me whenever I do a good job
- Keeps me regularly informed on important issues
- Provides me with ongoing feedback on my work

VALUES DIMENSION

Ethics / Personal Characteristics

- Leads by example and action
- Is someone I can trust and talk with openly without the fear of reprisal
- Keeps promises and stands by commitments made
- Acts decisively and gets things done

Justice / Fairness

- Treats me with respect
- Resolves conflicts fairly and appropriately
- Treats everyone equally—does not play favourites
- Recommends new ideas from our work unit up to senior management
- Gives credit to the whole work unit when receiving positive feedback on our performance
- Is fair when reviewing my performance
- Takes appropriate action with people who under-perform

In addition, the statements, "Provides me with ongoing feedback on my work" and "Recognizes or praises me whenever I do a good job," could be placed under either job performance or communication/feedback. We placed these statements under the communication/feedback category, because statements dealing with communication as a whole have high loading factors, and this also applied to the statements in the Leadership modules: the statements dealing with communication had a greater impact on employee engagement than the statements that dealt with straight performance issues.

Tier One Statements: Leadership

The following eight statements had the greatest impact on defining an employee's working experience and employee engagement:

- Leads by example and action
- Treats me with respect
- Is someone I can trust and talk with openly without the fear of reprisal
- Keeps promises and stands by commitments made
- Acts decisively and gets things done
- Resolves conflicts fairly and appropriately
- Listens with an open mind
- Treats everyone equally—does not play favorites

If we refer to our Four Dimensions of Leadership model, we see that seven of the eight statements fall under the values dimension. One statement in this tier one list, "Listens with an open mind," falls under the functional dimension. Four of the seven statements under the values dimension fall under the ethics/personal characteristics category, and three fall under the justice/fairness category.

The importance of these findings cannot be understated. Much of the published research shows the significance of questions in the functional dimension as being the most important contributors to employee commitment, retention, and overall employee engagement. For example, questions dealing with personal development, ongoing feedback, and recognition rank near the top. Our research shows that these are important contributors to employee engagement as well. However, our research also shows that statements from the values dimension rank higher than the statements from the functional dimension. In other words, a leader's personal characteristics of integrity, trust, and fairness play a greater role in employee engagement than do the functional behaviors.

Over the past few years, there have been numerous references to the notion that employees leave their bosses and not the organization. Our research paints a much clearer picture of the specific attributes of a manager who will engender employee loyalty.

These insights have direct implications for senior management and human resource professionals, who are charged with the responsibility of following up with appropriate leadership training. Standard management training is inadequate. When the subjects of integrity, trust, and fairness are broached, we are entering the realm of values that are deeply rooted in a person. Value-based training and coaching presents a major challenge for organizations. Values determine behavior. To succeed with behavioral change, one must address values.

Tier Two Statements: Leadership

Although we are calling this a tier two list, it should be noted that the loading factors are still strong (i.e., important contributors to employee engagement). The important point that is revealed by our research is that the value-based behaviors of managers play a more prominent role in employee engagement. The following eleven statements have load factors greater than (.700), whereas in the tier one list, the load factors are greater than (.800).

- Recognizes or praises me whenever I do a good job
- Recommends new ideas from our team up to senior management
- Gives credit to the whole team when receiving positive feedback on our performance
- Gives me the latitude that I need to do my job to the best of my abilities
- Ensures that I have the right skills and knowledge to perform my job
- Keeps me regularly informed on important issues
- Provides me with ongoing feedback on my job performance
- Supports me to develop myself for future advancement opportunities
- Provides me with clear performance expectations
- Is fair when reviewing my performance
- Takes appropriate action with people who under-perform

The tier two list collects all of the performance-related statements and picks up the remainder of the statements from the justice/fairness category. As one moves down the list, there are comparisons between similar statements that deserve a second look. For example, recognition for a job well done is more important than ongoing feedback on job performance. Having latitude to do one's job ranks higher than being provided with clear performance expectations. These examples give us insights into the characteristics of engaged employees. Engaged employees, by definition, are emotionally committed to their jobs and to their organizations. They are adults and expect to be treated as adults. As such, they need a degree of freedom to do their jobs. It is equally important for employees to feel that their manager trusts them as it is for them to trust their manager. This shows a basic tenet of an engaged employee: the relationship between supervisor and employee is based on an adult-to-adult relationship.

This two-way trust is seen in the second and third statements as well—"Recommends new ideas from our team up to senior management" and "Gives credit to the whole team when receiving positive feedback on our performance." The manager does not hoard the glory for good ideas and for a great job accomplished by her team. She is not afraid to acknowledge the good work of her team. She is not afraid to put forward useful ideas to her superiors from her subordinates. She recognizes that having a positive and fruitful relationship with her team needs to be based on trust. However, the manager is responsible for creating the trust work environment and has to take the first step. The manager has to demonstrate to her staff that she trusts them.

As for the fourth statement—"Gives me the latitude that I need to do my job to the best of my abilities"—there's a wonderful advertisement on television for a job-posting Web site that illustrates it perfectly. A man is meeting with his boss. He outlines a problem. His boss agrees with him that it has to be fixed. The man asks her, "What are you going to do about it?" She looks at him and answers, "Nothing." He looks puzzled. She continues, "You are going to fix it. You know more about this than I do." The tag line for the advertisement is "Are you looking for a good boss …"

The first four statements are followed by the statement dealing with skills and knowledge—"Ensures that I have the right skills and knowledge to perform my job." On reflection, this should not be a surprise. Once a trusting relationship between two adults is established, where the employee is given latitude to perform his work, having the right skills and knowledge is essential. This enables the employee to succeed, and it ensures that the organization prospers.

The last statement on the list, addressing under-performers—"Takes appropriate action with people who under-perform"—typically scores low with all of our clients. Most managers don't like dealing with conflict and will avoid it if they can. They will especially avoid a perceived confrontation with an employee that is under-performing. Our client work supports this assertion. We found the same applies to the question in tier one that deals with the resolution of conflicts—"Resolves conflicts fairly and appropriately." When we discussed this observation, our clients consistently stated that they found it uncomfortable dealing with employees who under-perform. However, when we discuss this issue with union leaders, we are always surprised by their response. They feel managers should deal with under-performers because it is unfair to the other employees if a coworker is not pulling his weight.

A final interesting fact to note about this list is a statement that didn't appear: "Reviews my job performance at least once a year." This statement was part of our employee surveys—we have subsequently dropped it—and it is part of all employee surveys we have seen offered by other firms. Our research leads us to the conclusion that recognizing an employee and providing feedback on an ongoing basis are far more important contributors to employee engagement than reviewing a person's job performance once a year. There is additional information in Chapter 8, "Notes on Employee Surveys," that discusses the pitfalls of time-based questions in employee surveys.

Leadership Statements Linking to My Department

The next analysis we conducted involved looking at the correlates between the statements in the Leadership module and the statements in the My Department module. At the broadest level, the Leadership module correlated with Mission and Values most strongly, followed by Corporate Practices and My Department. This came as a surprise. However, as we sifted through the data, it became apparent that management, first and foremost, puts their stamp on the mission and values of the organization. They define who the organization is and what it stands for. They do this in words, but more importantly, they show it through their actions.

The same can be said for corporate practices. All of the statements in the Corporate module are totally under the control of senior management. In essence, the Corporate module contains statements that put into action the principles as they are set out in the Mission and Values module. There were fourteen statements that linked strongly with the My Department module:

- Resolves conflicts fairly and appropriately
- Keeps promises and stands by commitments made
- Gives me the latitude that I need to do my job to the best of my abilities
- Treats everyone equally—does not play favorites
- Ensures that I have the right skills and knowledge to perform my job
- Acts decisively and gets things done
- Takes appropriate action with people who under-perform
- Gives credit to the whole team when receiving positive feedback on our performance
- Supports me to develop myself for future advancement opportunities

In this analysis, six statements come from the values dimension and three from the functional dimension. As in the previous section, the values dimension is still predominant over the functional. However, three behaviors dealing with job performance load strongly with departmental practices: latitude in work, having the skills and knowledge, and having development opportunities. The My Department module of our employee engagement survey defines the immediate working environment of an employee. In light of this, it is not surprising that these three statements from the functional dimension of our Leadership module load strongly with departmental practices. Having latitude and skills to do one's work, and being supported for personal development, extend the description of one's immediate working environment.

All the published research suggests that the ranking of questions as they apply to employee engagement is the same for every organization. Our research shows that this is not the case. We found that the importance of leadership in determining the level of employee engagement varies from one business sector to another. One size does not fit everyone. In addition, we discovered minor variances in ranking within the leadership statements themselves. Typically, the same statements would appear in the tier one statements, but they might rank in a different order from one company to the next. These results come from over forty-five thousand employees in our database.

Let's look at two organizations: one, an insurance company, and the other, a healthcare organization. The healthcare agency ranked "leading by example and action," "taking appropriate action with people who under-perform," and "recommending new ideas up the line" as their three most important leadership behaviors. At the insurance company, "leading by example and action," "resolving conflicts fairly and appropriately," "taking appropriate action with people who under-perform," and "giving the latitude needed to do the job" all ranked equally in importance.

Drilling further down, we discover that "giving encouragement to do the job the best [I] can" and "giving credit to the whole team when receiving praise" were next in importance for the healthcare agency. At the insurance company, the next set of behaviors was: "provides me with clear performance expectations" and "recommends new ideas from our team up to senior management." We noticed these small variances from one organization to the next. However, it was clear that out of the twenty-two statements in our Leadership module, questions from the value dimension predominated.

Other variances emerged from sector to sector. The most striking was the difference between a high-tech and a retail company. Leadership behavior did not play as predominant a role in the high-tech firm. In fact, Mission and Values and Corporate Practices ranked higher than Leadership and My Department in the high-tech sector. In contrast, Leadership ranked head and shoulders above the other survey modules in retail.

Over a period of four years, we discovered that the leadership scores of store managers were an excellent predictive tool of store performance. A retail company with over two hundred stores and ten thousand employees routinely promoted high-performing store managers from a smaller store to a larger one. Or it promoted assistant store managers to store managers. Some high-performing store managers were rotated to troubled stores. There were many other reasons for moving a high-performing manager from one store to another. This movement of managers provided an ideal laboratory to measure the impact a high-performing manager has on the business success of a specific store. This company identified high-performing managers by the leadership scores a manager received from our employee engagement survey and by their business measure. Some of the business measures included sales per square foot and secret shopper scores.

We tracked the movement of all managers, as measured by our survey, from one year to the next. Our survey report contained a table that ranked all of the stores in descending order from the highest to lowest survey scores. This table showed the overall survey score as well as a column for the score of each module in the survey. In this case, the modules were: My Store (instead of My Department) My Manager, Corporate Practices, Mission and Values, and Stress.

As we tracked the movement of store managers with high leadership scores (direct report scores), we noticed that a pattern developed. For example, one year a particular store was ranked number forty out of 212. However, the store manager's score was very high. The following year, this same store was ranked number one. We discovered that, in the previous year, the store manager had just been promoted to that store and had not had a chance to make his imprint. We looked at our data and the stores' internal measures more closely. We discovered that there was a strong correlation between the leadership scores of store managers and the store ranking, not only based on our survey scores, but also based on sales per square foot, secret shopper scores, and lower rates of employee absenteeism.

When these findings were presented to senior management, they began to invest considerably more in training store managers. This new training included modules in behavior. When we started to work with this regional retailer, their profitability ranked in the middle of their four thousand stores worldwide. By 2004, the 212 stores in Canada had the highest profitability in all three brands on a worldwide basis. This is an example of a company that had powerful information and, by acting on it, made a huge difference to their bottom line.

The difference we found in the importance of leadership on employee engagement, between the retail sector and the high-tech sector, lies in the nature of the work itself. The nature of work in retail is heavily dependent on people contact: manager to employees, employees to employees, and employees to customers. The nature of work in high-tech is light on people contact and heavy on technical work done alone in front of a computer. Contact with one's manager is infrequent. Employees periodically participate in project-based team meetings, but the bulk of their time is spent alone working on their computer, being creative, and looking for solutions. It is not surprising that leadership plays a smaller role in employee engagement in the high-tech sector.

This example shows that not every company is created equally. We disagree with the survey firms who offer the same set of questions as measures of employee engagement to every organization in every business sector. Our work with clients and our research clearly show that there are differences among business sectors, and that one size does not fit all.

In the case of Ontario Hydro, the impact of leadership behaviors listed in the tier one list fell somewhere in the middle, between the retail and the high-tech examples. The property agents worked alone for the majority of their time. As we discussed in an earlier chapter, their job design was flawed. However, whenever they had contact with their immediate supervisor or the head office manager, it was a negative experience. This negative experience fueled the fires of an already unhealthy work environment. The lack of respect, trust, recognition, and support from their local and head office managers added to their de-motivation and poor health.

In trying to think of some personal examples (other than Northern Energy) from my past, where I was dramatically impacted by my manager's behaviors, one in particular comes to mind. Two years after I graduated, I was working for the federal government as manager of a new department responsible for federal property development. The early seventies was a period of an unprecedented surge in property values, followed by a period of high inflation. As an owner of vast properties across the country, and in recognition of the rapidly increasing land values, the federal government launched the Federal Land Management Policy in the mid-seventies.

The policy had two parts. In recognition of the escalating property values, the first part of the policy directed that all federal lands be put to their highest and best use. With the aim of maximizing federal land use, the second part identified a significant decentralization of federal departments out of the capital to regions across the country. The Atlantic east coast was economically depressed and the recipient of annual equalization payments. In light of this, they were assigned several large projects. The significance of these projects was amplified tenfold because they were being located in small communities, providing an infusion of temporary construction employment as well as hundreds, and sometimes thousands, of new full-time jobs.

My department had the responsibility of conducting environmental impact studies for these projects, and working with the local provincial and municipal governments to ensure that the projects were developed in conjunction with local zoning. The magnitude of these projects typically required zoning bylaw changes. My department was responsible for acquiring the necessary changes. While this was taking place, we were also in the midst of conducting highest-and-best-use studies of several large tracts of surplus federal lands that involved public hearings after our studies were completed. My team comprised bright young graduates with backgrounds in urban and regional planning or economics. My boss, Bob, was a career civil servant who had tenure, but who had no education or experience in property development or environmental impact studies. This was not an unusual situation in the federal government at that time.

We were getting ready for a presentation to senior officials from all three levels of government. My team and I worked day and night for several weeks on the proposals for this development. A couple of days before the meeting, the team and I met with our director and spent several hours briefing him on the project. We went through all of our overheads—no PowerPoint in those days—several times. We were led to believe that the members of the team who worked on the project would accompany me to the meeting. An hour before the meeting, Bob came to my office to say that he wanted me alone to accompany him to the presentation, since this was a gathering of high-level officials. After the customary introductions, to my surprise, Bob began the presentation. After the presentation was completed, Bob received positive feedback and congratulations on a job well done. He was happy to receive all the accolades and did not acknowledge the hard work of my team. I was devastated. I found the experience hurtful and demotivating.

When I reflect on this incident in light of the Four Dimensions of Leadership, Bob's actions fell squarely in the values dimension. Bob did not treat my team or me with respect. He did not give us credit for a job well done. He could not be trusted, and he did not exemplify high standards of honesty and integrity. Although I did not analyze the incident at the time, I intuitively knew I could not work for a person like Bob. The job, the opportunity, and my team were fantastic, but the incident motivated me to look for an exit strategy.

Managing Self First

Leadership is a combination of strategy and character. If you must be without one, be without the strategy.

—H. Norman Schwarzkopf

I once heard a story of a young priest who had been working as an assistant to a bishop. The bishop was old and got very sick. He was lying in a hospital bed with his young assistant by his side. The assistant had become attached to the bishop and respected him for the wisdom he had shown. Knowing that the bishop was rapidly approaching the end, the young priest asked the bishop if he would share the most important lesson he had learned over the course of his long ministry. The bishop obliged and said the following:

> When I graduated, I was really fired up. I could see no limits to the possibility for making a significant contribution. I wanted to change the world. A little while later, I received my first charge, a small parish in a rural community. I was still fired up. I was going to make a real mark. I joined a couple of diocesan committees, and I developed a plan for all the changes I was going to make in my little parish.

After a while, I began to experience a great deal of frustration, with my committee work and with my parishioners. The committee work moved at a snail's pace. My parishioners were resisting some of the improvements that I wanted to introduce. They did not want to move as quickly as I did. Nevertheless, I achieved some things and, after seven years, I was moved to a much larger church.

I decided to lower my expectations. Clearly, I was not going to be able to change the world. Perhaps I could make a significant contribution to the diocese. I worked hard. I experienced more frustration. Introducing change at the diocese was a formidable task. Obviously, my work did not go unnoticed. After a number of years, I was asked to be the regional dean. Okay, I thought. Now I'll work hard and I'll really make a mark on this deanery. But my efforts to change the deanery for the better ran into all kinds of obstacles.

Many more years passed and I was elected bishop. I surveyed my charge and thought that maybe there were one or two parishes that I could help to improve. And my efforts were also met with resistance. As I lie here in this bed, thinking about my ministry and my life, I realize that I was going about it all wrong. I was starting at the wrong end. If I was going to make a difference in this world, I should have started with myself. I should have focused on improving and changing me.

I now realize that the greatest contribution any one of us can make, especially if we have a leadership role, is to work at becoming the person that God wants us to become. It is important for us to be Christ-like. We have to be masters of our life, first and foremost. Our life as a living example of the spirit of God will have a much greater impact on making this world a better place.

This little story illustrates a profound truth about the importance of managing self first. If a leader cannot manage himself, how can he expect to lead others? A leader needs to be a role model first. A leader's behavior speaks louder than his words. It is not surprising to see that the value dimension plays such an important role in effective leadership. A manager cannot be an effective leader until she knows who she is. She cannot manage effectively unless she:

- Knows her strengths and limitations

- Is physically fit

- Is mentally fit and spiritually rooted (i.e., her life is built on a foundation of solid values)

- Is driven by a clear mission or higher purpose for her life that transcends her personal self-interest

Knowing Yourself

One of the marks of great leaders is that they recognize that they don't have all the skills that are needed to succeed. They recognize the power of a team. They build a team that will complement their strengths and compensate for their limitations. Research has shown that the characteristics of good leaders don't follow the stereotypical picture of a larger-than-life leader with a big ego who is seeking the limelight. In fact, they show the opposite. For example, in his book *Good to Great*, Jim Collins gives the example of Darwin Smith, the CEO of Kimberly-Clark. Mr. Smith is described as a man who blended extreme personal humility with intense professional will. Collins found this type of leader at every one of the good to great companies.

Collins called these CEOs Level 5 leaders. He defined them as leaders who channeled their ego needs away from themselves and into the larger goal of building a great company. He says, it's not that the Level 5 leaders have no ego or self-interest. They are ambitious, but their ambition is, first and foremost, for the company and not for themselves.

In keeping with building the right team, Collins used the analogy of a bus. Level 5 leaders start their job by getting the right people on the bus and getting the wrong people off the bus. They build a strong team first. They also always give the team credit for successes achieved and take the blame for failures. Only a person who really knows himself can do that and not feel threatened.

Physical Fitness

It is not the purpose of this book to discuss physical fitness. However, a discussion of leadership is incomplete without acknowledging the importance of physical fitness and proper diet. Good physical health promotes high levels of energy. There is a large body of research detailing the benefits of regular exercise to a healthy cardiovascular system, strengthening the heart, restoring bone density, and helping to cope with stress. One cannot underplay the importance of the saying, "If you look after your body, your body will look after you." Stress is a constant companion of great leaders.

Leaders have to make tough decisions. They have to perform at peak levels all the time. With their book *The Power of Full Engagement*, Jim Loehr and Tony Schwartz helped many athletes excel at their sport. They noted that top athletes train 90 percent of the time in order to perform 10 percent of the time. However, business leaders have to perform at peak levels ten, twelve, or more hours every day. The physical demands on today's leaders are huge. If they don't have the physical stamina, they'll become fatigued and their judgment will be impaired.

To be an effective leader, physical fitness should become a daily routine, a natural, almost unconscious, part of a leader's life. It must be fully integrated into a leader's life, just like brushing their teeth. It should be something they do daily without giving it much thought.

Mental and Spiritual Fitness

I have combined the discussion of mental and spiritual fitness because I cannot talk about one with talking about the other. To be mentally tough, one has to be spiritually tough. I am a person of faith. I begin each day with Bible reading, prayer, and meditation. These daily spiritual exercises are no different than daily physical exercises. They are fully integrated into the daily routine of my life. I do them—period. In the same way that my morning workout prepares my body for the events of the day, my spiritual exercises prepare my mind. The spiritual exercises remind me of who I am, what my life represents, what my values are, and what my purpose in life is.

When I complete my physical, mental, and spiritual exercises, I know that I will be the best that I can be. I know I'll have a good day. I know that I'll be better equipped to handle whatever the day throws at me. This doesn't mean I won't make mistakes. Sometimes, I may not make the best decisions. However, I do know that I'll make better decisions following my daily routine of mental and spiritual exercises than I would otherwise make. I know that I'll be grounded in my values and ethics.

A person who has value confusion, or what I like to refer to as "situational values" and "situational ethics," cannot be expected to provide effective leadership. This person will not be someone most will want to follow, because his or her subordinates will feel they are standing on shifting sand. Robert Greenleaf's book, *Servant Leadership*, best demonstrates this value-based understanding of leadership.

Greenleaf writes: *"The servant-leader is servant first. Becoming a servant-leader begins with the natural feeling that one wants to serve, to serve first. The conscious choice brings one to aspire to lead. That person is sharply different from one who is leader first, perhaps of the need to assuage an unusual power drive or to acquire material possessions. The leader-first and the servant-first are two extreme types. Between them are the shadings and blends that are part of the infinite variety of human nature."*

Being a servant-leader is based on a value of service or servanthood. This value will shape a servant-leader's decision making, interaction with peers and subordinates, and overall behavior. In his book *Leading Your Self,* Jagdish Parikh says that unless someone knows how to lead himself, it would be presumptuous to think that he can lead others. For a leader to be effective, the leader has to be clear on his or her values, believe in those values, and be solidly rooted in them. If I asked you to name a leader you admire, you might name the Dalai Lama because he represents the highest standards of ethics and wisdom. Those are the hallmarks of a great person and a great leader.

Parikh suggests that a manager who aspires to become an effective leader must experience a higher level of self-awareness that goes beyond one's ego. This is true, but it is also a huge challenge for anyone. I believe it is easier to attain within the context of a faith community, where the values are clearly articulated and reinforced every week. I also believe it is attainable in secularism. Michael Josephson, an ethicist, offers a clear secular approach to ethics and ethical behavior. Josephson has written numerous books on ethics and is the founder of the Josephson Institute of Ethics in Los Angeles, California. Josephson says that there is no such thing as personal or business ethics—there is only ethics. Many people prefer flexibility. They are attracted to situational ethics. But Josephson makes it clear that ethical behavior and decisions based on ethics are black and white.

Ethics are manifest in what we do and what we say. The challenge for leaders is to ensure that what they do and what they say are completely aligned. This is difficult to achieve without a deep knowledge of oneself, periodic self-examination, and being deeply rooted and committed to a set of personal principles of conduct. Josephson lists twelve ethical principles: honesty, integrity, promise keeping, loyalty, fairness, trust, concern for others, respect, law-abiding, commitment to excellence, being a role model, and accountability. These principles provide a comprehensive list for a leader to adopt and to exhibit in everyday decision making and in personal and business relationships.

A Word about Trust

Although Josephson provides a comprehensive list of ethical principles, most of these principles flow out of trust. Only one question in the leadership set of questions in our survey specifically contains the word "trust"—"Is someone I can trust and talk with openly without the fear of reprisal." However, most of the questions flow out of trust and have a direct bearing on trust. Research has shown that trust and fairness are two values that expand an employee's capacity to deal with stress and difficulty in the workplace.

"Productivity and trust go hand in hand," William Ouchi wrote in his book *Theory Z.* The level of trust that exists in an organization, more than any other factor, is the most important element in determining high performance. Trust creates an environment where people feel secure. It creates an environment where employees can reach their maximum capability. Communication opens up and creative ideas flow freely throughout the organization. Where trust is strong, barriers break down. Trust is essential to creating a seamless organization. It creates a cooperative environment for achieving maximum synergy.

If employees feel that they can trust their managers, that they can trust the organization, they become bonded and committed to the organization. If employees recognize that trust is a major part of the belief system of the organization, the employees are involved in a closer and more productive relationship with management and with each other.

Trust as a fundamental value predetermines the attitude toward all relationships and, specifically, the negotiation stance between union and management, supervisor and employee, employee and customer, employee and suppliers, president and the board of directors, president and lenders and investors. Trust, or lack of trust, is the key determining factor in shaping an organization's total attitude. Is it an attitude that will foster positive and productive relationships, or is it a destructive attitude that will act against the organization? In a negotiation situation, will the organization seek a win/win or win/lose posture? These are all determined by the degree of commitment the organization has toward trust as a core corporate value.

A major study at the University of Western Ontario, involving ten thousand public servants and forty thousand private sector employees, by Linda Duxbury of the School of business at Carleton University and Chris Higgins at the Richard Ivey School of Business, showed that employees who worked for supervisors who trusted them and respected them, reported less stress and greater productivity than employees who worked for managers who demonstrated little trust toward employees and who created stress. They coined the term "supportive supervisor" for the supervisors who engaged in two-way communication with their subordinates, provided ongoing recognition and positive feedback, allowed employees autonomy, and facilitated the completion of tasks.

Their research showed that there was a direct correlation between the bottom-line results of a company and the type of managers that predominated. They went on to say that employees with supportive managers were more likely to have high job satisfaction, high organizational commitment, and lower levels of job and life stress. They were also more likely than those with non-supportive managers to feel that the organizational policies of their company were supportive of them. This is particularly interesting because the organizational policies were the same for both groups—it was only the way these policies were implemented that was different. Finally, the data suggested that employees who worked for supportive rather than non-supportive managers were also more likely to feel secure about their jobs, trust their managers, and freely provide suggestions and feedback up the line. These outcomes are critical success factors for organizations that want to successfully implement change, re-engineering, and organizational restructuring.

Just as trust builds employee commitment, it also builds customer loyalty. Do your customers trust you to provide them with a consistent level of product or service quality? Do they trust you to back up your products with exceptional service? Do the customers trust you enough to come back again and again, despite attractive offers from the competition?

Because trust is so fundamental to defining the nature of relationships, it has a direct impact on the nature of communication in the organization. In a high performance organization, managers and employees are good listeners. Employees are encouraged to communicate their suggestions and ideas up the line, and these ideas are seriously considered, with fast and effective feedback provided. Communication is open when, for example, performance feedback is sought from employees about their supervisors through an enabling and constructive process. An atmosphere of trusting communication is encouraged to stimulate creativity and an open exchange of ideas. When problems arise, supervisors assemble their staff and open lines of communication to find solutions, rather than lay blame. There is a willingness to consider opposing views, and people are not put down or made to feel inferior if their ideas are not accepted.

In their book *Built to Last*, Porras and Collins show that visionary companies don't strive for comfort. When you are constantly challenged to be better and to push the envelope of self-improvement and excellence, the comfort zone is a never-attainable target. Porras and Collins found significant evidence that the visionary company has a history of tangible "mechanisms of discomfort" that impact change and improvement from within, before the external environment demands change and improvement. However, discomfort as an agent of positive growth can only occur in an environment of trust. If trust is absent, an uncomfortable work environment is destructive, and innovation and improvement will not take place.

Building Trust

Understanding is the next step to building trust. It is difficult to trust a person, product, or organization without the benefit of sufficient information or understanding. Consistency and predictability are two key elements in developing trust.

- Is the person true to their word?
- Is the product reliable?
- Does the organization put customers first?
- How consistent and predictable are the answers to these questions?

Information and understanding about the person, product, and organization are needed to answer those questions. This knowledge is gained through a process of research and data collection. On a personal level, the knowledge is typically obtained through communication between two individuals. This involves asking questions, listening to the answers, and responding to questions asked.

The best results and greatest degree of understanding will occur in an atmosphere that is free of judgment, where two individuals can engage in a dialogue as equals. A common communication model presents three positions from which a person can communicate. These are parent, child, and adult. When a person assumes the parent position, they automatically "hook" the other person into a child position and vice versa. When a person assumes the adult position, it naturally "hooks" the other person into an adult position.

The nature of communication transpiring from a "parent-child" positioning is filled with judgment, discomfort, control, and manipulation. This type of communication inhibits the free flow of information. Typically, the information flows in one direction and solicits a reaction by the receiver. The originator then reacts to the receiver's response with another dictum, and this is repeated several times until the interchange is ended by the "parent."

When a person assumes an adult position and sets up an "adult to adult" dialogue, there is no judgment and no control. Removing judgment creates a comfortable and secure atmosphere. The conversation is characterized by an open flow of questions and answers. Both parties listen, generally without interruption. This type of dialogue fosters understanding and eventually builds trust between the individuals involved.

In the workplace, one-on-one conversation predominates, either between coworkers or between the employee and their team leader. As a manager, the team leader has the power to set up either an "adult-adult" communication position or a "parent-child" position. In management language, a comparison can be made to Douglas McGregor's "Theory X" and "Theory Y." Simply stated, Theory X is a dictatorial management style where employees are told what to do and communication is unidirectional, going from management to employee. Theory Y represents a participatory management style where communication flows two ways, between management and the employees. This style sets up the adult-to-adult communication relationship. McGregor goes on to show that performance of Theory Y companies is superior to that of Theory X companies.

In his book *Maverick*, Ricardo Semler was one of the first to demonstrate the power of independent work teams, and the power of enabling employees to reach their full potential by treating them as adults. One of the contributing factors to the success of the teams was the open communication that flowed between the co-workers to solve problems and to improve productivity.

Despite the positive results gained with independent work teams, a large majority of organizations still operate out of the traditional model of line departments. Various forms of management by objectives predominate. The annual performance review is still a common management practice. The very nature of the performance review automatically sets up a parent-child communication relationship, where the supervisor is in judgment of the employee. This invites stress and anxiety and does not invite open and constructive communication. These models do not operate on trust.

A Purpose Driven Life

The last aspect of self-management is having a clear mission or higher purpose for your life that transcends your personal self-interest. Rick Warren wrote a compelling book called *A Purpose Driven Life*. Although Warren's book is rooted in his Christian faith, it nevertheless describes the importance of looking outside one's self to search for a higher purpose in life. Warren succinctly writes, "It's not about you." The mistake most of us make is to look to ourselves to find our purpose. We fixate on our needs. Do I want power, fame, fortune, or all three? But the answer does not lie in these questions. Rick Warren argues that the answer starts with God. What is it that God wants for your life? What is God's purpose for your life?

Loehr and Schwartz don't refer to God. From their perspective, the source of your purpose is spiritual. It is the spiritual energy that is derived from our values and a purpose beyond our self-interest. It is not my intent to get into a religious or faith argument in this book. Finding one's purpose can be difficult at the best of times. However, from my perspective, it's easier to look to God for answers and to adopt Greenleaf's proposition that to lead is to serve. What servanthood looks like will be different for every leader, depending on their personal life situation. The key premise is that leaders must have a clear picture of their higher purpose to be truly great leaders. This provides the right perspective for leadership.

Again, Darren Smith of Kimberly-Clark is a good example. His purpose was to make his company great. It was not to make him great. It was not to make him famous. It was to make the company great. Mary Kay built her cosmetics empire on a philosophy that, from the newest recruit to the chairman of the board, everyone had to live by the Golden Rule—Do unto others as you would have them do unto you—and three priorities: God first, family second, and career third. She called this the "go-give spirit" and believed it was the reason behind her company's success. By giving of oneself and being sincerely interested in the total person, we are able to bring out the best in others and in ourselves.

As we move through life's journey, our purpose will change. I recently read an article about Darren Entwistel, President and CEO of Telus Corporation, the second largest telecommunications provider in Canada. Darren is only forty-four years old. He has achieved a major transformation of Telus in his six years as CEO. However, he has publicly declared that when he is finished at Telus in four or five years, he will walk away from corporate life. Darren says that when he looks at his personal balance sheet, the return on investment is hard to justify. He has paid a high personal price for his achievements. He is a man who could write his own ticket in the corporate world, but he is ready to give it all up to become a professor of strategy and leadership. Darren's purpose will change.

Don Edwin's purpose was fueled by his desire for personal power. He seemed only to be focused on self-interest. Don is highly intelligent. He's a strategic thinker and he knows how to make things happen. However, he did not care about his organization or the people working in the organization. His focus was on feeding his own ego drive for power and recognition. There was little evidence that he cared about his customers, who were viewed as a vehicle for profit. There was a strong connection between Don's leadership style, his leadership behaviors, and the toxic nature of the company. Although we might at first think that Don's behavior was the point of greatest toxicity, if we look at the data from our research, we can see that the most long-term destruction was caused by the negative working conditions.

Effective self-management is an ongoing process. You cannot stop when you reach a particular point. Physical, mental, and spiritual fitness have one thing in common. If you stop your exercises when you reach a certain point, you will slowly lose what you have already achieved. Fitness has to be maintained.

Several years ago, I identified ten individuals who were either presidents or vice presidents in ten different organizations. (I drew from a pool of approximately three hundred senior business contacts that I had made over the years.) Several, but not all, were my clients. I was impressed with these senior managers because they possessed six qualities. They were:

1. Strategic thinkers. They knew how to think outside the box, to plan in a way that was best for their customers, their employees, their local community, and their organizations.

2. Systems thinkers. They were very much aware of how a particular action in one area of the business would impact another part of the business.

3. People of action. They planned strategically, and they executed their plans successfully. They knew how to make things happen.

4. Highly intelligent. Most had an MBA. In addition to the MBA, five had a doctorate degree (in various disciplines), one also had a medical degree, and another had a degree in veterinary medicine.

5. People who cared about their employees and their organizations. They had strong people skills and were enabling leaders. They worked hard at creating and running great organizations.

6. People with a strong sense of purpose. Some had a very strong Christian faith, while others where profoundly spiritual. They were driven by a purpose higher than their own self-interest.

I organized a dinner for these ten individuals. The purpose of the dinner was to introduce them to each other and to see if, after they met, they would like to continue to meet in the future, to learn from each other, help each other solve problems, and generally support each other. It's lonely at the top, and this would provide them with a safe forum for discussion.

Half the group was women. This was not by design but because of who they were. All ten knew themselves very well and were solidly grounded in their own set of values. They knew their strengths and weaknesses. They were mentally and spiritually fit. They had a clear purpose and direction for their lives. The only trait that was varied was physical fitness. Some exercised regularly, while others did not.

The dinner was a huge success. We continued to meet regularly for about two years. As it turned out, the meetings were timely for many of the folks. We learned a great deal from each other. We helped each other solve problems. Some problems were strategic in nature as they related to their businesses. Others concerned organizational and personnel issues. In fact, a large chunk of our time was spent on organizational and personnel issues. In addition, discussing personal issues took up a lot of our time together.

As you can imagine, these individuals were in high demand. Several were offered new positions, either inside or outside their organizations. We helped a number of members sort out these offers and helped them to reach difficult decisions regarding their personal career moves. One accepted a position that took her to London, England. Another CEO decided to leave her organization and become an independent consultant. Another was wooed by a huge conglomerate. He was offered double his current salary; however, the opportunity required a move that would have been extremely disruptive to his family. After a great deal of discussion, he declined the offer. The meetings were timely.

We disbanded because at least five changed jobs and moved. They accepted more interesting and more challenging assignments. We were dispersed and, as it is written in Ecclesiastes 3:1, "For everything there is a season, and a time for every matter under heaven." The group served its purpose and we moved on. All ten members are great leaders, and it was a privilege to share the two years with them. They are exceptional people, working in small, medium, and very large multi-national companies, about which you will never read. But they are blooming where they are planted, and they are making a significant difference. To this day we are good friends and keep in touch.

Chapter 5

Mission and Values

Where there is no vision the people perish.

—Proverbs 29:18

Many companies have their vision and mission statements posted in the lobby of their building. Some of these statements are long and others short and to the point. Some are accompanied by statements of principles.

Surprisingly, they all tend to look the same. You can't tell the difference between one company and another. In fact, in many cases, you can't tell what products or services the company is offering. All of them seem to contain variations of the same phrases along the following lines:

Our vision is

• To become the best ...

Our mission is

• We are the leading ...

Our guiding principles are:

• We value our customers
• We value our employees
• We work to create a profit and ... shareholders ... etc.

Most of these statements are vague and virtually meaningless. You could mix up various companies' vision and mission statements and no one would be able to tell the difference. Few of them have the power to mobilize the energy of their employees. They are worded in such a way that they miss the purpose of creating a vision and mission statement in the first place.

Vision and mission statements should be worded in a way that all management and employees can understand, remember, like, and be passionate about and committed to their organization's mission, core purpose, and reason for being. They know how their work fits into these statements. They know where the team is going long term (vision). Everyone knows and shares the same values and beliefs.

Therefore, in a high performance organization, everyone knows what their individual contribution is to the overall purpose of the organization. They know why they are there and what is really important. They now exactly how they fit in; why their function is important; what they have to do; and how they have to do it. They know exactly what behaviors are expected of them. They know exactly to what standards of performance they must work, and to what level of quality and excellence they must aspire. They know exactly how to behave and interact with their team members. But above and beyond that, they want to behave in the prescribed way, and they are totally committed to the values and culture of the organization. Their personal and corporate values and beliefs are the same.

Within the racing team, there are sub-teams. The designers work closely with the engineers and the drivers. They, in turn, work closely with the mechanics and the pit crew. Some teams have long-term performance objectives; some have very short-term performance objectives. For example, the contribution of the pit crew is measured in seconds. Four seconds to change tires and six seconds to refuel. The mechanics' timeframe is a bit longer during the race, should the car experience a "fixable" mechanical failure. The work of the designers and engineers, at this stage, is relegated to monitoring and assessing.

Mission Statements

Over the last few years, most companies have adopted a mission statement. Some did it because they recognized the importance of the mission as a "driver" for the organization. It is the definitive statement of what the organization is and what it stands for. It represents the heart and soul of the organization and gives meaning and purpose to the organization. It goes beyond describing the organization's products and services. The mission statement breathes life and meaning into the organization. It is the highest order of expression of the organization and, therefore, does not change over time. It is an expression of the purpose to which the company has been called. Goals, objectives, and strategies will change over time, but the mission will not.

When I worked at Ontario Hydro, the senior management hired a consultant, who helped them create a vision and mission statement and guiding principles. The result was a thirty-six-page booklet that each senior manager carried in his pocket. I recall being at a meeting with a number of senior executives on a discussion about electricity pricing, and one of the executives pulled out this little booklet. He had a funny smile on his face and said, "Let's see what it says about pricing in our bible." It was the first time I had heard about the booklet. Subsequently, the vision, mission, and guiding principles were distributed to everyone as a memo. I don't recall having any formal presentation or discussion with my boss at a managers' meeting. So I decided there was no point in discussing it with my staff either. It was obvious that it was not a priority. In retrospect, I now see that this was an excellent example of how *not* to do a vision and mission statement.

- The process was backwards. Everything was done by senior management without the involvement of frontline staff.

- The moment the statements were committed to print, they became irrelevant. A thirty-six-page booklet that was carried in people's pockets, instead of their minds and hearts, completely missed the mark.

- The communication process, or lack of one, was the certain death knell of the vision and mission. No one paid any attention to it.

Many years ago, Entec developed a simple and effective process for creating a meaningful and energizing vision and mission. We used this method with several clients to great success. When we completed the Organizational Health Survey with one of our clients, it became apparent that the client did not have a mission statement. At the subsequent action-planning workshop that comprised twelve frontline employees and twelve managers from every management level and every region, there were four tables of six people. When we reached the part of the workshop agenda set aside for developing their mission statement, each table was asked to draw one picture as a team. They were given one piece of paper from a flip chart, magic markers, and crayons. They had one hour to draw what they thought their company's mission would look like.

At the end of the hour, each team presented their picture and described it to the group. As they described their pictures, I wrote key descriptive words on a flip chart. The pictures were amazing. The first group presented a picture of a large circle of people holding hands. They weren't just holding hands in the normal fashion. They were holding hands in an interlocked fashion, where each person crossed his or her arms across the front of their body. All of the figures were beautifully dressed in brightly colored clothing. The group described what they had drawn. They saw:

- People as part of a community
- People as friends
- People having fun
- Happiness and joy
- Good looks
- Great clothes

After the applause for the first group abated, the next group stood up. They tacked their picture on the wall, and I got goose bumps looking at it. Conceptually, their picture was identical to that of the previous group. They also had drawn a large colorful circle that looked like a beautiful wreath, from a distance. Upon closer examination, the circle was made up of people holding hands. There were people of every color, and all were brightly dressed. When all the presentations were completed, the key words that described the pictures were essentially the same as that of the first group.

We considered these words as a group and took just fifteen minutes to come up with the following mission statement:

"To make people feel good by selling them great clothes in a fun and happy environment."

Their mission statement was simple, easy to remember, had emotion, and spoke to the hearts and minds of the employees. It was presented to senior management after the workshop and adopted.

The statement subsequently drove many decisions that helped to create a working experience for the employees that truly reflected the mission statement. Along with this and the other work that we did with the company to raise the level of employee engagement, it's interesting to note that, within four years, the two hundred Canadian stores in this company of four thousand stores and a hundred and eighty-five thousand employees worldwide, became the most profitable in the world.

This success was achieved by a major retailer because they took the results and recommendations of our Organizational Health Survey (now our Level 2

Employee Engagement Survey) and acted on them. They now had a clear mission that provided the fuel and imagination for everything they did. It drove senior management decision making. It drove their training. It drove the leadership behaviors of all the store managers. The store managers were responsible for creating an amazing working experience for their staff, in which the staff could have fun and make people feel good by selling them great apparel.

The following examples show a range of mission statements from good to poor:

Walt Disney

Purpose

To make people happy

The Walt Disney purpose statement is an example of a great statement. It is short, easy to remember, and meaningful. It is focused outside of the company and on their customers. This statement has stood the test of time.

CIBC

Our Goal

To be the pre-eminent Canadian financial services company

Our Vision

Winning customer loyalty through service excellence

Our Values

- *Commitment to stewardship*

- *Respect for every individual*

- *Encouragement of initiative and creativity*

- *Excellence in everything we do*

The CIBC example shows a reasonably clear and concise progression from goal to vision to values. It may not be as powerful as the Disney example, but each statement is reasonably short and easy to remember.

The next example comes from a company whose mission and vision is of medium length, neither short nor long.

AT Plastics

Mission Statement

*To be a profitable, environmentally responsible, high-quality specialty plastics pro-
ducer, deploying advanced technologies and innovative solutions to meets the needs
of our valued customers*

Our Vision

*To meet the needs of our customers in North America and internationally with
innovative and continuously improving high value plastics specialty products*

The last example comes from **Ericsson Communication**. Written several
years ago, it is lengthy. Their mission and vision are contained in a six-page booklet.

Our Mission

*Our mission is to understand our customers' opportunities and needs and to pro-
vide communication solutions better than any competitor*

This is followed by two explanatory paragraphs.

Our Vision

To be the choice for:

- *telecommunications solutions that unleash customer potential*

- *personal development opportunities that unlock employee potential*

- *unparalleled technological excellence in all telecom products and services*

This is followed by another paragraph explaining the meaning of this in
greater detail.

Our Shared Values

Professionalism, respect and perseverance

Followed by three paragraphs, each one with a lengthy description.

Our Guiding Principles

- *We will listen actively to our customers and jointly define success*

- *We will look for ways to work together to find total business solutions*

- *We will support, recognize, and reward innovation and risk taking*

- *We will support and reinforce personal and professional development*

- *We will listen actively to each other and jointly define expectations*

- *We will fulfill all commitments—we will do what we say we will do*

- *We will celebrate our success*

- *We will turn failure into continuous improvement*

- *We will act with honesty, integrity, and professionalism*

- *We will pursue opportunities for mutual gain in all relationships*

- *We will define and deliver solutions in an intelligent and cost-effective manner*

As we continue to protect our leadership position and enjoy the triumphs of even greater achievements, we will use our guiding principles as our—"rules of the road"—the code for the spirit and behavior which defines the way we achieve each goal on the way to our vision. With all your help and commitment, we will make the vision real.

This section is followed by eleven explanatory paragraphs.

These examples show a spectrum of approaches used by companies to document why they are in business, and why their employees should be motivated to work with all their energy and effort.

Consider these vision and mission statements and ask the following questions:

- Which of these statements will be remembered by the employees?

- Which of them will the employees be able to remember for future reference?

- Which will excite and motivate the employees to give as much as they can to their organization?

- Which set of statements can the employees refer to when a problem arises and there are no specific guidelines or policies that they can reference to help them resolve the problem?

- Do the employees see how they personally contribute to the mission of the company?

- Do they see how their specific job function contributes to the mission?

Further questions should include the CEO or president:

- How often does the president refer to the mission in his or her speeches?
- Do the employees get excited and inspired when they hear the president referring to the company's mission in speeches?

Values

Values guide the choices we make. They determine how we act and, in situations where there are few guidelines, they determine the decisions we make. Values define the core of who we are as individuals, and they define the core of all organizations. Individuals as well as organizations can have a strong and well-defined set of values, or they can have a weak value base. Those with a strong value base stay the course and remain true to their core beliefs, even as circumstances around them change. Those who don't have a strong value base will change with the shifting wind, with changes in circumstances. In their book *Built to Last*, Collins and Porras discovered that the highly successful companies, or visionary companies, as they were referred to, all had a strong core ideology that did not change over time, even though the external environment kept changing. These companies were able to move forward, to reinvent themselves as was necessary to remain the top companies in their field, but their core ideology never changed.

Thomas J. Watson, Jr., former CEO of IBM, wrote in a booklet, *A Business and its Beliefs*, "If an organization is to meet the challenge of a changing world, it must be prepared to change everything about itself except its basic beliefs as it moves through corporate life. The only sacred cow in an organization should be its basic philosophy of doing business."

Many organizations have recognized the link between values and performance. They have discovered that their organization operates with a set of values that is not necessarily openly defined, but which nevertheless influences the way their employees think, act, and react. Although everyone arrives in an organization with their own set of values that has been formed over the years, if the values of the organization differ from their own, tension occurs. The extent of this tension will depend on how firmly an individual is entrenched within their own value system, and how different that value system is from that of the organization. If the person does not hold fast to any particular belief system, he or she will readily be able to adopt the values of the organization. If the values of the person and the organization are the same, there will be no tension, and the person will feel at home in the organization and fit right in from the very beginning.

However, if the person's values are vastly different than the values of the organization, and if the person is firmly committed to their values, they will find it difficult to adapt, and will feel uncomfortable working in the organization. Conflicts will arise, and there is a high probability that the person will quit or be fired. If they stay, they will be unhappy, and the emotional trauma and tension caused by the conflict may manifest itself in some form of mental and physical difficulty. This may show itself in a variety of ways: stress-related headaches, back pain, frequent colds, and flu.

Research Findings

Our research bears this out. At the broadest level, the Mission and Values module in our survey loads most strongly with the Departmental Practices module as noted in Chapter 3. In addition, Mission and Values has the strongest statistical link to Mental and Physical Energy.

Recognizing the importance of being intentional about naming their values, many organizations define the values they would like to see, and print and post these for all to see. This is fine so long as these values are being practiced on a daily basis by everyone in the organization. Problems arise when values are posted, but senior management does not set an example by living them out on a daily basis. The values of the president will influence those of his management team, which in turn will influence the rest of the organization. If these are different than what is posted, the organization will experience value confusion, and employees will find that they don't know how to act in vague situations.

Although companies have policies to guide employee behavior, these policies cannot always cover all situations. In those vague situations when an employee must make a decision, a clear set of values plays a key role. If the policy does not cover a particular incident, does the employee run to his supervisor for a decision? This is a frequent practice in many organizations that is both time-consuming and wasteful.

In order of importance, the following statements on values have the greatest impact on employee engagement:

- Everyone in my department is focused on quality improvement
- I have not been subjected to verbal abuse or any kind of bullying this past year

- Everyone in my department behaves with integrity, honesty, and fairness
- Creativity and innovation are valued at our company
- I am clear on how important my contribution is to the vision of our organization
- I am proud to work at this organization
- Our organization's mission excites me and motivates me to do my best work
- Everyone in my department treats each other equally, regardless of gender, color, or ethnic background

Reflections

At first we were surprised to see "focus on quality improvement" at the top of the list. However, after reviewing our other data, it became apparent that statements where employees are looking outside of themselves have a significant impact on both engagement and health (i.e., those that deal with the employee's desire to do a good job, to contribute to quality products and services). This statement is not "about me," (inward looking) but it is about making a contribution (looking outside of self).

However, the next two statements are "about me" and they define the values that employees consider to be important as they define the nature of working relationships. The fourth and fifth statements are outwardly focused.

Reflecting back on Ontario Hydro, all of these components that are representative of a healthy work environment were absent. Quality, integrity, and honesty were not top of mind. Ironically, fairness, as it related to land acquisition, was an espoused value. But in Hydro's efforts to appear fair, they distorted the boundaries of integrity and humanity.

There is not much to add to the story at Northern Energy. Abuse and bullying were the norm and not the exception. The fear that Don created spawned a culture of finger pointing and blame. Everyone was in self-preservation mode. No one wanted to make a mistake and be found at fault. This brought out the worst in most people, where malicious politics, gossip, and backstabbing thrived. Managers and staff were equally complicit in these activities.

Several years ago, we worked for a client who was a fundamentalist Christian. He claimed that he ran his company on Christian values. Every Friday morning, the senior management team met at a restaurant for an early breakfast, followed by prayer and Bible study, and concluded with a discussion of the company's progress. The company needed some assistance in dealing with growing pains, and the company president was referred to me through a mutual colleague.

When we began our work, I met many of the employees. All of the senior and middle managers were Christians. It was not evident at first, but after some time, it became apparent that the source of many of the problems in the company was a massive gap between what the president professed verbally and what he actually did. He never stopped looking for ways to cut corners. He especially focused on suppliers because he felt they were dispensable. He looked for ways to avoid paying his suppliers. He was involved in several court cases where suppliers sued for non-payment. The sums of money in these cases totaled over one million dollars.

In one case, the company received $800,000 worth of product and the president refused to pay. Cash flow was tight. He was approached by the supplier on a number of occasions and still refused to pay. The supplier sued for payment. The case was dismissed at trial on a technicality.

Several months after this case was dismissed, I had lunch with the company's lawyer, a friend of mine. I had completed my work for our mutual client, and I quizzed him about the other lawsuits the client was involved in. I discovered that my friend had no knowledge of his client's double standard. The lawyer took the information his client gave him in good faith and defended him to the best of his abilities. And he won every case. He always found some loophole that got the president off the hook. He saved the company over a million dollars. Near the end of lunch, my friend confided to me that he had not been paid. He was owed $45,000 in legal fees, and he asked me if he should sue the president for payment.

The gap between what the president said and what he did resulted in a state of paralysis in the company. Senior managers and middle managers were incapable of making decisions because of the value confusion. No one knew where they stood. Fear prevailed, and decisions could only be made after consulting with the president. When I had completed the business-related work for the company, I decided that I needed to discuss the values gap with the president. I was aware of the personal risk. We had been paid for our work on an ongoing basis, but there was still a significant portion outstanding. To my surprise, the president was quite interested in what I had to say. He genuinely seemed to take my feedback to heart.

Not long after our talk, I learned that the president asked the chief financial officer to change the financial statements that were being prepared for the past financial year. The CFO told the president that what he was asking was illegal. The president said he didn't care and insisted that the statements be changed.

The CFO resigned. I understand that a couple of years later, the president filed for bankruptcy.

This example raises a number of questions:

- What was the relationship between the stated values of the organization and the actions of the president?

- Although he professed a high standard of values, and surrounded himself with like-minded people, what kind of individuals remained in the company?

- What was the impact of this president on his business relationships: employees, suppliers, and banker?

- What was the impact of the value confusion that pervaded the company on decision making and achieving business success?

- What was the impact of his actions on the perception of the Christian faith?

Suggestions

- When it comes to values, the less you talk, the better. Let your actions speak for you.

- Let your employees create your mission statement. You will be surprised; they know your business better than you do.

- Keep your mission statement simple so that your employees can carry it around in their hearts and minds.

- Integrity rules. All business relationships—employees, suppliers, investors, community, and other stakeholders—must be based on the highest standards of integrity.

Chapter 6

The Work/Health Connection

Before we begin to discuss the connections between organizational practices and leadership behaviors with employee well-being, let's briefly review the Employee Engagement Model. This model is depicted below. Four components influence employee engagement: health, leadership, mission and values, and organizational practices and processes. The Individual Health module is divided into two parts, because there are factors over which an individual has control (lifestyle), and external elements over which they have little or no control (death of a loved one).

Similarly, within the organizational context, there are practices over which an employee has varying degrees of control. Most employees have some level of input or say regarding practices at the team or departmental level. Typically, they have little to no say regarding corporate-wide policies and practices regarding governance, mission, and business strategy.

Employee Engagement Model©

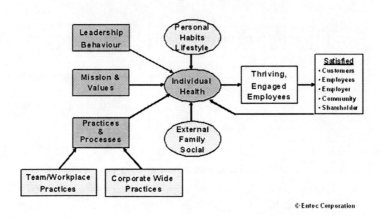

© Entec Corporation

The model provides the logic for the selection and placement of questions. The Individual Health portion of the survey measures mood (symptoms typically associated with depression), mental and physical energy, (symptoms typically associated with stress and burnout) and attentiveness (symptoms typically associated with disorders that affect our ability to focus or to pay attention). A fourth module was added to the well-being module: ability to perform. This module comprises three questions:

1. Am I able to carry out all of my work responsibilities most of the time to the best of my abilities?

2. Am I able to nurture my close and intimate relationships?

3. Am I able to maintain a full and satisfying social life?

If I have two glasses of wine, I'm done. I cannot go back to my office and work. Yet there are people who can have two glasses of wine, and can return to their office and do their work with minimal impact on their performance. Similarly, different people have different levels of coping skills. One person can complete the "mood" section of the survey and receive a "high risk" score, and yet indicate that they are able to carry out all of their work responsibilities. Another person will receive a "high risk" score and indicate that they cannot carry out their work responsibilities.

After we completed a number of survey projects, we conducted correlation analyses using Pearson Coefficients, a commonly used correlation methodology. These projects included organizations in different sectors and involved over five thousand employees. Two overall findings were evident:

1. There was a strong correlation among the four organizational modules: departmental practices, leadership behavior, corporate practices, and mission and values loaded strongly with each other. (Departmental practices and mission and values linked with each other most strongly, followed by departmental practices and corporate practices.)

2. There was a strong correlation among the four well-being modules: mood, mental and physical energy, attentiveness, and ability to perform. (Mood and mental and physical energy loaded onto each other most strongly, followed by energy and ability to perform.)

Additional Research Insights

1. Organizational scores and well-being scores moved in tandem. As the organizational scores increased (improved), the well-being scores increased (improved) as well. As the organizational scores deteriorated, so did the well-being scores.

2. There was a strong link between mental and physical energy and three organizational components, namely, departmental practices, corporate practices, and mission and values.

3. There was a moderate (subordinate) link between mood and departmental practices, and mission and values.

4. There was also a moderate (subordinate) link between leadership and mental and physical energy.

5. No link was found between leadership and mood or attentiveness.

6. The data showed that mood was a leading indicator. In other words, of the three well-being factors being measured, there were always more employees at high risk for a mood disorder, followed by mental and physical energy, and attentiveness.

7. The high correlation between the three well-being modules would indicate that a person with a mood disorder was also afflicted with a

depleted state of mental and physical energy (burnout) and diminished attentiveness.

To illustrate these observations graphically, the following model was developed:

Organization & Employee Wellness Interface Model©

Organizational Components   Employee Well-Being

Entec Corporation©

Implications

Since mental and physical energy is the only wellness module that strongly links with the organizational components, we can conclude that:

1. The measure of an organization's contribution to the emotional well-being of employees can be determined by the number of employees who are at high risk or who are already suffering from a depleted state of mental and physical energy.

2. The number of employees who are suffering from a mood disorder also includes those who are suffering from a depleted state of mental and physical energy and attentiveness. Therefore, the increment between the number of employees who are experiencing a depleted state of mental and physical energy, and the number who are suffering from a mood disorder and affected attentiveness, can be used to show the number of

employees whose well-being can most likely be attributed to personal factors that are external to the workplace.

The implication of this finding is that there are two categories of employees in the workplace: those who are hurting as a result of pressures in the workplace, and those who are suffering for personal reasons. If either should reach a point where they are unable to function, and goes on short- or long-term disability, the design of the return-to-work process needs to recognize the source of the disability.

Impact of Mission and Values

When we began to mine our database, we realized that we had enough information for another book. Therefore, in this book, we are going to limit our discussion to the overall relationship between the organizational and health modules in the survey, and on the relationships among the individual questions. We will also show the individual well-being questions that link with the four organizational modules. These analyses will provide a fresh look at practices, leadership behaviors, and values that appear to have a greater impact on employee well-being and that impact employee performance.

At the highest level, the mission and values module had the strongest link with the health modules. This module linked most strongly with mental and physical energy, with a lesser link to mood and ability to perform. Departmental practices had a slightly weaker loading than the mission and values module. Corporate practices followed. Leadership loaded weakly with mental and physical energy.

These results surprised us. Intuitively, we thought that departmental practices or leadership would link most strongly with the health modules. However, this was not the case. It appears that leadership impacts health indirectly, as it is reflected through organizational practices and organizational values. In addition, we were surprised to discover that mission and values played the most dominant role.

In analyzing individual organizational questions that linked with wellness, we looked at frequency (the number of times an individual question linked with a particular wellness module), the strength of the correlate, and if there was a link with ability to perform. These analyses surfaced questions that had the greatest impact on employee wellness. The following is a list of questions, in order of significance, that linked to mental and physical energy:

1.   Our organization's mission excites me and motivates me to do my best work*†‡

2.   I am clear on how important my contribution is to the mission of our organization*†‡

3.   Everyone in my department is focused on quality improvement

4.   Everyone in my department works in a spirit of collaboration, rather than confrontation or competition

5.   I am proud to work for this organization*†‡

6.   Everyone in my department behaves with integrity, honesty, and fairness

7.   I know what my customers/clients want, and I am focused on their needs in my work*†‡

8.   Creativity and innovation are valued in our company

9.   Our senior managers frequently talk about our mission

* In addition, this statement links to an employee's ability to perform.
† In addition, this statement links to an employee's mood.
‡ In addition, this statement links to an employee's attentiveness/ability to focus

Reflections

It is striking to see that the first three statements refer to mission and quality improvement. This again reinforces the importance of the mission in the lives of employees. The focus on quality improvement seems to flow naturally out of a compelling mission. There is another important insight. The top three statements deal with the employees looking outside of themselves. They deal with the employee's excitement about doing a good job and making a contribution, and the employee's contribution to quality products and services. These statements that link to employee well-being are the same as those that were among the strongest contributors to employee engagement. These statements are not "about me," (inward looking) but about contribution (looking outside of self), and being part of a higher purpose.

The link between all the statements on mission and employee health supports our research findings with clients, where we conducted a correlation analysis between the rate of absenteeism and an employee's knowledge of the company mission. The employees with the lowest rates of absenteeism knew their company's mission statement. The employees with the highest rates of absenteeism did not know their company's mission. This finding was a breakthrough in demonstrating the important contribution that can be made by a compelling mission that is clearly communicated and known by employees.

All the statements that link to employee health are the same as the statements that contribute to employee engagement, except for statements 4 and 7, above. Statement 4—"Everyone in my department works in a spirit of collaboration rather than confrontation or competition"—describes the nature of the working relationships among coworkers, and this link is repeated in a similar statement—"My coworkers work well together"—under department practices. Statement 7—"I know what my customers/clients want, and I am focused on their needs in my work"—is related to the family of statements that address mission. This statement is also outward looking in terms of being able to meet customer needs. Again, we cannot underscore enough the importance that employees attribute to being able to do their jobs unencumbered: knowing where the company is heading, and knowing what their role is in helping the company to succeed. Our data demonstrates that this promotes good employee health and employee engagement.

At Ontario Hydro, there was no mission statement. We were building nuclear plants and transmission lines. There was no higher perspective or calling to the work. (As I have stated, this was later corrected by senior management. They introduced a 36-page booklet that contained the mission and guiding principles. I recall the vice presidents carrying the booklet in their shirt pocket for quick reference. No one could commit to memory the points in the booklet.)

The agents knew they had a deadline that they had to meet. They were not driven by a higher purpose. Quality was no longer relevant. This deadline was the only driving force. The property agents were not proud of their organization. Integrity, honesty, fairness, collaboration, and quality were not on the radar screen. Innovation, however, was valued in the context of creating a "fairy tale valuation," as one property agent described what his job entailed.

We conducted research with a client—one of the largest retailers in the world—and discovered that there was a direct correlation between an employee's knowledge of the company's mission statement and the rate of absenteeism. Employees who knew the company's mission statement had the lowest rate of absenteeism, and the employees who did not know the company's mission had the highest rate of absenteeism. This was a valuable insight, because it demonstrates that a good mission statement that is known has a quantifiable impact on an organization.

This current research further confirms that being excited by the mission statement and knowing how one contributes to the mission links to a person's well-being and ability to perform. More specifically, we now know that these two statements link to mood, which measures the systems of depression, and to attentiveness, which measures a person's ability to focus and concentrate on their work. Finally, being excited by the mission, and knowing how an employee contributes to the mission, affects the employee's ability to perform their work. From our perspective, this is a huge finding, because managers at all levels in most companies do not attribute much significance to their mission. They merely give it lip service and may refer to it in passing.

Impact of Departmental Practices

Departmental practices follow closely behind mission and values in terms of their link to employee well-being. The individual statements that linked to mental and physical energy, in descending order, are:

1. I feel that I am a success at work*†‡
2. Information is shared openly by my colleagues in my department
3. My coworkers pull together to find a solution in a time of crisis
4. I find my work challenging and stimulating*†
5. I am able to contribute what I do best every day*
6. I have easy access to the resources, people, and information I need to do my job
7. I rarely experience conflicts with the managers I work with
8. I can complete my work within the time allotted without pressure most of the time*†
9. My coworkers work well together
10. The scheduling of my work is fair and reasonable*

11. I can manage the number of interruptions, email messages, and telephone calls I receive each day *

* In addition, this statement links to an employee's ability to perform.
† In addition, this statement links to an employee's mood.
‡ In addition, this statement links to an employee's attentiveness/ability to
focus

Reflections

Comparing the above statements to the statements in Chapter 3, "Organizational
Practices," that had the highest loading with employee engagement, we can see
that there are differences in the factors that impact employee health, and factors
that contribute to employee engagement. Practices that are related to "time" link
with health but not engagement: finding solutions in a time of crisis, completing
work without the pressure of time, and managing interruptions. The only statement that does not fit is, "I find my work challenging and stimulating." In this
case, an employee's level of interest in their work is a determinant of their level of
mental and physical energy. When a person is motivated, their level of energy
rises. However, our data shows that having challenging and stimulating work is
not in itself enough to fully engage employees. A workplace community that is
based on cooperation, fairness, and justice is essential for employee engagement.

When I first arrived at Ontario Hydro, my staff was experiencing all of these
statements in their negative form. In addition to reducing mental and physical
energy, our research shows six of the practices link with an employee's ability to
perform: statements 1, 4, 5, 8, 10, and 11 directly impact an employee's ability to
perform. In their negative form, they reduce the employee's mental and physical
energy to the point where it significantly impairs the employee's ability to do
their work.

Four of the statements, 1, 4, 8, and 11, also link to mood, which measures the
symptoms of depression. When you put all of these negative practices together,
you have all the elements of a high-demand and low-control work environment.
You have the ideal conditions for high levels of stress. It is not surprising to see
that there was one death, two individuals who suffered strokes, and a high level of
alcoholism in the property acquisition department at Ontario Hydro.

Considering these statements from the vantage of my personal experience at
Northern Energy and the boss from hell, I can see that all the statements that
linked to mental and physical energy were present in the negative. Even though I

was succeeding, by his interruptions and frequent tirades, Don managed to make me feel insecure and unsuccessful. When I did achieve the ultimate success—completing the highly successful marketing program—I was fired. I was always under pressure of time and could not work at my best every day. The interruptions were constant. The work was challenging, but after a while, I began to lose my passion for the work. In retrospect, I can now see how the seeds of my depression were sown.

Impact of Corporate Practices

The following statements linked with mental and physical energy. In descending order, they are:

1. Our incentive and reward system motivates me to do my best work

2. Candidates that are best suited for a job are promoted when job openings become available

3. I am provided with ample opportunities for personal development

4. My organization provides the right amount of information [for my needs] on its direction and performance

5. I can cope with the pace of change at our company*†

6. I know how changes at our company will affect me personally

7. My organization is reasonable in allowing me to balance my work with my personal life†

8. The initiatives offered by my organization are appropriate for my health and wellness needs†

* In addition, this statement links to an employee's ability to perform.
† In addition, this statement links to an employee's mood.

Reflections

The practices that link with health deal with fairness, personal development, communication, coping with change and, in the case of the last two statements, employee health considerations. Three statements with asterisks link with symptoms of depression. The statement that addresses one's ability to cope with change also links with an employee's ability to perform their job. The absence of these practices creates an atmosphere of uncertainty. The uncertainty leads to anxiety and stress. In the case of corporate practices, the practices that are ranked the highest for employee engagement are the same as those that link with employee well-being.

In the context of Ontario Hydro, all eight practices were absent. The property agents did not feel fairly treated, and there was no thought to personal development or to issues of work/life balance or personal health. Corporate communication was poor to non-existent as it related to company changes and to the impact those changes may have had on employees.

Similarly at Northern Energy, it was difficult to cope with the pace of change. There was no thought given to work/life balance and the gym was not sufficient for my health and wellness needs to compensate for the working conditions. When Don Edwin made the organizational change that introduced the high performance team, he left a lot of questions unanswered. In fact, he created a new layer of confusion and added to my anxiety by refusing to talk about their role and responsibilities. I was given no information on how the high performance team was supposed to work with the marketing function. Fairness was not on Don's radar screen. Don handed out promotions on a whim. They were subjective. All of this added to my anxiety and stress.

Impact of Leadership Behavior

Although the Leadership module as a whole did not statistically link with any of the health modules, there were several individual statements that linked with mental energy. They are listed below. My immediate supervisor:

1. Gives me the latitude that I need to do my job to the best of my abilities

2. Treats everyone equally—does not play favorites

3. Keeps promises and stands by commitments made

4. Ensures that I have the right skills and knowledge to perform my job

5. Gives credit to the whole team when receiving positive feedback on our performance

6. Keeps me regularly informed on important issues

7. Recommends new ideas from our team up to senior management

8. Provides me with clear performance expectations

Reflections

Based on my personal experience with Don Edwin, I found it surprising that the Leadership module, containing twenty-one statements on leadership behavior, did not link with either mood or attentiveness/mental focus, and weakly linked with mental and physical energy. This is despite the fact that leadership behaviors have a significant impact on employee engagement. We can conclude that the impact of leadership on employee well-being manifests itself through best practices and processes at the corporate and departmental levels. A senior manager's skills in implementing best practices throughout the broader organization, and with local people managers, has a much greater impact on the health of employees than the quirks of personality.

Digging deeper, we first looked at the leadership statements in this section from the perspective of the four quadrants of our Leadership Model that was presented in Chapter 4. In this case, there were three statements from the job performance and three from the ethics/justice quadrants. One statement came from the communication and one from the personal characteristics quadrants. In total, there were four statements from the functional and four from the values dimension. This is significantly different from the eight behaviors, shown in Chapter 3, that had the greatest impact on employee engagement. The tier one list contained leadership behaviors that had the greatest influence over employee engagement. In that list, seven of eight were from the values dimension and only one was from the functional dimension.

However, in this case, there are only two statements from the tier one list (2 and 3), both of which are from the values dimension. The rest are from the tier two list; they are evenly split between the functional and values dimensions.

This leads us to believe that there is a more balanced weighting between leadership behaviors from the functional and the values dimensions impacting employee mental and physical energy. We had to reconcile this difference between employee engagement and employee well-being.

We looked to Ontario Hydro for answers. The heart attack, strokes, and alcoholism were not caused by my predecessor's lack of values. His lack of values, as defined by the values dimension in our Leadership Model, acted as a de-motivator, but this was not enough in itself to cause ill health. However, combining the lack of values with no control over their job, and contradictory performance expectations, (statements 1 and 8, above), proved to be a deadly combination.

Let's consider Don Edwin for a moment. His personal behavior and leadership style were inexcusable. However, it was his complete disregard for orderly practices and processes that created the most havoc. His interference in day-to-day activities and his micro-management effectively seized any control we had over our work. There were no clear performance expectations. These were like a moving target. Northern Energy was his personal domain. He treated it as his plaything. Best practices were non-existent. Don's "practices of the moment" ruled the day. Northern Energy is a good example of the balance that is needed between job performance and ethics/justice for reducing stress and promoting good health.

One Final Thought

Many organizations have embraced a series of health promotion programs. Senior managers need to be cognizant of the importance of fixing the organization first. Too many organizations think that by introducing programs for stress management, weight reduction, quitting smoking, blood pressure clinics, and lunch and learns, they have done their part for their organization. However, misaligned practices and unsupportive leadership behaviors that are embedded throughout the organizations undo any good that the health promotion brings. The organization must be fixed first, in order for employees to gain the full benefit of available health promotion programs.

Chapter 7

A Successful Employee Survey Process

An employee survey should be an integral part of an overall management process. The survey should not be an isolated event. There are eight key points to conducting a successful employee survey:

1. Senior management commitment

2. Clear objectives regarding post-survey outcomes

3. Pre-survey communication process

4. Simple survey questions that can be understood by all levels of education and multiple ethnic groups

5. Clear survey analysis that leads directly to follow-up implementation

6. Post-survey communication process

7. Post-survey action planning and implementation

8. Post-implementation communication

The Communication Process

Several years ago, a company president approached Entec Corporation to conduct an employee engagement survey. I explained to him that our pre-survey communication process includes a memo from the company president, promising to follow up with substantive post-survey improvement actions. He responded by saying he would decide to follow up only after he had seen the survey results. I apologized and said that we could not do business together unless he provided his commitment to follow up on the survey results before the survey. He said he wouldn't and I politely excused myself.

A senior management team should not undertake an employee survey unless they are prepared to commit to follow-up action. They are wasting their money, and they are fueling the fires of employee cynicism and discontent. By committing to follow-up action, the senior management team is allocating the survey within a management process that is designed to improve the performance of the organization. Entec provides our clients with four pre-survey communication templates. Each communication builds on the one that preceded it. All the memos are from the president. The first communication is aimed at directors and managers. It introduces the survey and details the objectives of the survey. It includes the president's commitment to follow-up action.

The second communication is exactly the same as the first, except this time it is sent to all employees. Both of the memos note the survey launch date. The third communication template focuses on protecting employee privacy and confidentiality. It specifically details all the steps that are taken to protect employee privacy and confidentiality. The memo addresses issues concerning the minimum number of respondents needed in a team for scoring purposes. It describes the firewall that exists between the organization and Entec Corporation, so that no one can access an employee's individual data. The fourth communication is a summary of all the points: senior management commitment to follow-up action, survey objectives, and confidentiality. The last memo is issued a couple of days before launch. When the survey process is completed, another memo is sent to all employees, thanking them for their participation.

After the survey report is presented to the senior management team, there are several follow-up options the client can use to move forward with implementation. These vary from organization to organization. They depend on the culture of the organization and the survey results. For example, if the survey results are reasonably positive, the follow-up process will be simpler and shorter than if the results are poor and the organization requires extensive post-survey work.

Our surveys are organized in a way that clearly groups questions into two categories: actions over which local people managers have control, and actions over which senior management has control. The following chart depicts a generic process for follow-up action planning and implementation that reflects this separation in control. This process can change from one organization to another to reflect organizational differences.

EMPLOYEE ENGAGEMENT

ACTION PLANNING & IMPLEMENTATION PROCESS

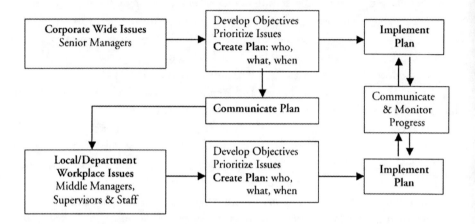

In most cases, Entec is asked to facilitate this process on behalf of a client. This helps the client to focus on developing the proper action plans without concerning itself with the mechanics of running the process.

Action-planning Workshops

One option that is a favorite of first-time clients is a two-day action-planning workshop. The ideal number of participants is between twenty-four and thirty. The pool of twenty-four would comprise twelve frontline employees and twelve managers from every level. The twelve frontline employees are chosen from every division or department, or, in the case of an organization with multiple geographic locations, there would be an employee from every region or office location. In this case, managers are also chosen from each geographic location.

Since the pool of managers has to be from every management level, there may be a supervisor from one office, a manager from another, a director from another, and so on. We call this a diagonal slice through the organization, because a different level manager is selected from each region or department. The diagonal slice accomplishes two things: each region or department is represented by a member of management, and every level of management is represented. Employees from head office are also represented. The vice president of human resources is always present. In addition, if there are managers or individuals in an organizational development or strategic planning role, they are asked to participate.

In a company with several thousand employees, the challenge that has to be addressed is the fair selection of workshop participants. To deal with the equity issue, we ask our clients to run a contest. Anyone who wishes to participate in the action-planning workshop has to submit an application. The application is simple. It includes a line for the name of the employee, department/location, job function, and a space for the applicant to describe, in fifty words or less, why they want to participate in the workshop. The vice president of human resources reads through all the applications and chooses the participants.

Over time, this selection process has become standard. The workshop is considered a prize by employees because it is typically held at an off-site location, such as a golf course or resort. Some companies hold the workshops in a hotel near their office.

The workshop starts with dinner on the first day. After dinner, the president will briefly talk about where the company is heading, and will create a context to impress upon the workshop participants the importance of the workshop and its outcome. This is an important aspect of the workshop because the company president is seen as the sponsor of the workshop. After dinner, we hold an "issues bazaar," where employees sign up to work on four to six issues. The number of issues depends on the total number of issues that were identified by the survey. When an issue has six signatures, it is no longer available to anyone else. This process ensures that each participant will work with a different set of colleagues throughout the workshop.

The participants work in groups of six, at four or five separate tables. Each table will work on a different issue, allowing participants to work on four or five issues. One and a half hours is given to each issue. The groups work through a set of questions, such as why the issue exists in the first place, barriers to fixing the issue, what it would take to remove the barriers, a proposed action plan identifying specifically what resources are needed, what needs to be done by whom, and within what time frame.

At the end of the ninety minutes, each table presents their results to the rest of the group, at which time the whole group can comment on and add to that issue. When the workshop nears its end and all of the action plans are posted, a champion is selected who will take full responsibility for making sure each action plan is implemented. Nominations are taken, and the workshop participants vote for the champions. The workshop recommendations are presented to the senior management team by the elected champions. Priorities are finalized at this meeting. In addition, if new resources are needed, they will be approved by senior management at this meeting, or they will be taken under advisement and a decision provided within two weeks.

The role of the champions has proven to be a key factor in the success of follow-up implementation. Depending on the size of the organization, one member of the senior management team will be the champion for the overall process. This could be the president, vice president human resources, or vice president strategic development or organizational development.

Another option is a senior management team workshop that can last anywhere from three hours to a full day. The focus of the senior management workshop is to design the follow-up action-planning process, post-survey communication, and the issues that senior management will undertake.

The Post-survey Process

Organizations with whom we work a second or third time will run with the development of post-survey action plans. For example, in the retail sector, each store manager is provided with their store scores, and the store manager is given responsibility for developing an action plan. We assist the company with a template to be used by the store manager. In another example, an insurance company, every department manager is given their department's scores. The department manager and his or her staff prepare a follow-up action plan.

In all cases, the action plans that are prepared are approved by senior management. The store manager or department managers are held accountable for the successful implementation of the plan. In one particular company, the store manager's performance bonus is based on her success in achieving the objectives in the action plan. Also, part of her bonus is based on the leadership score she receives on the annual employee engagement survey.

In the third survey cycle, our approach with our clients changes. At this point, they have had two years to initiate a number of necessary organizational improvements. These could include leadership training, streamlining of departmental practices, improving both local communication and corporate communication, and introducing health promotion programs that emerged from the survey. To make substantive gains in the following year, we shift gears. This shift in gears is necessary because over time a pattern develops with all of our clients. The same issues emerge year after year, but the severity of the issue is not as grave. In other words, the follow-up actions that our clients initiated were not wasted. They yielded improvements, but the issues were still not resolved.

The shift that is taken in the third and subsequent years is to focus on the client's strengths and to build on those strengths. This could be as simple as getting greater leverage from their high-performance managers. This leverage can take two forms. The first is to provide increased support to the high-performing managers, and to build a stronger and larger department with expanded responsibilities. Another approach involves rotating a high-performing manager into an under-performing or troubled area of the organization. Both of these approaches have proven to be effective. Choosing one or the other depends on the unique circumstances of the organization. In some sense, this is what happened with my property acquisition role at Hydro, even though it was not the conscious intent when I was hired. Personally, I prefer the first approach, but there are circumstances when the second approach is preferable.

A well-organized post-survey process is key to achieving the greatest benefit from an employee survey. The examples given above are just a couple of samples from a larger number of different approaches. The approach has to be tailored to the unique culture and situation of the organization. For example, the process we use in a non-union organization is different from the process we use with a unionized organization. In a unionized organization, both the pre- and post-survey processes are different. We have separate sessions with the union leadership before the survey is launched, so the union membership understands the purpose of the survey, and also understands the benefits to union members. Similarly, after the survey is concluded, we have separate sessions with the union leadership, where we share and discuss the results with the union leaders. Also, in the action-planning workshops, we always ensure that a sufficient number of union members are represented.

When the various implementation teams successfully implement their initiatives, they must communicate this to the rest of the organization. We worked with a colleague that did an excellent job of implementing various initiatives after our action-planning workshops. Eighteen months later, we discovered that the senior management was criticized by the unions for doing nothing with the survey results. They did an excellent job, but they didn't broadcast the good works they were doing. There is a natural tendency to do this, but letting the organization know the good works that are being undertaken is an important activity. This client learned the importance of ongoing communication.

At the other end of the spectrum, another client did an excellent job of communicating to staff. They launched a new leadership behavioral coaching program. As part of the launch, they went to great lengths to communicate the fact that the new program was a result of the employee engagement survey. They were intentional in carefully communicating the link between all new initiatives and the employee engagement survey. I cannot emphasize enough the importance of communication in the post-survey implementation phase.

Chapter 8

Notes on Employee Surveys

Many organizations conduct employee surveys of various types, either annually, every two years, or sporadically.

Every organization has a different reason for conducting an employee survey. Over the years, I've met company presidents who think that an employee survey is a good thing to do, but they haven't thought about what they would do with the data once they got it.

I have also met senior management teams whose primary motive for running an employee survey in their organization is to compare their scores to the scores of other companies, and see if they can become one of the "Top 100 Best Managed Companies," or some other publicly visible "awards" program. They view the employee survey as a marketing or public relations initiative. They don't consider the negative impact on their employees of conducting a survey and doing nothing with the results.

Many surveying companies sell their services on the basis that they will be able to compare the scores of a company against the other organizations in their database. Comparing yourself to someone else is enticing. We have all been exposed to comparative data from the first day we stepped inside a school. Throughout our primary and secondary education, we were compared to everyone else, based on the "class average." We knew who the smartest and the dumbest kids were, but we wanted to know if we were above or below the class average. It was important in terms of our own self-esteem and dealing with our parents' expectations. The parents of some students demanded top marks, and that is what those few students worked toward. They had to be the best. They had to have the top marks.

One could argue that comparing ourselves to other organizations is, in fact, a legitimate strategic objective. It is worth knowing how you compare to the best. How does your stock performance compare to the best in your business sector—not the average of all the companies in your business sector, but only the best? How do your employee survey scores compare to the best in your business sector—not the average of all the businesses in the database, but only the best?

Comparing ourselves to the very best is legitimate, especially if the best sets a benchmark that we adopt as our own. But to compare ourselves to the average serves no useful purpose. If a senior management group knows that their scores are better than the average of all the companies in a database, what strategic use is this information? It might give them a sense of pride, knowing that they're better than the average, but it might also lull them into a false sense of confidence. The questions that should be top of mind are: Are we really as good as we can be, and are we really achieving a level of excellence that will sustain us over the long term? How will we use the survey information to improve our organization? How will we use the survey information to improve the working experience of our employees?

For example, employee turnover in the retail sector is fairly high. Most retailers take it for granted. But it doesn't have to be that way. Earlier in the book, we discussed how one retailer was routinely raided by other retailers. Their annual turnover rate for store managers was 39 percent, and for associate managers 48 percent. That was average for this sector. However, the turnover rate was costing the company $1,560,000 per year in recruiting and training. With over two hundred stores and ten thousand employees, these costs were unacceptable. After Entec worked with the company, the turnover rate of store managers was reduced to 13 percent in one year. The net annual savings of this reduction was $1,040,000.

Plus, these lower turnover rates were accompanied by real business gains. For example, secret shopper scores increased by 5 percent after only eight months, and sales in Canada over the last few years have improved to a level where the Canadian operation moved from being somewhere in the middle, to becoming one of the most profitable divisions in the world. The survey results were linked directly to the bottom line.

If this company had accepted "the trap of comparing themselves to the average," and accepted the conventional wisdom that "this is the average turnover rate in retail, so we're okay," they would not have saved $1,040,000 each year in training and recruiting. More importantly, they would not have experienced the benefits that reduced turnover brought them, namely, preserving human capital of highly trained managers that helped to grow their business. This last point is typically overlooked. The impact of a reduction in turnover of well-trained employees to the bottom line of a company is considerably higher than the cost savings achieved by reducing recruiting and training.

Others I have met decided to run an employee survey because they wanted employee feedback. They wanted to "test the temperature of the water," one president said to me. When I asked him if he would do anything with the results, he responded by saying that it all depended on the results. If the results were good, he would share them with his employees. If they were poor, he would probably file the survey report. He had no intention of following up with any action.

In all of these examples, the organizations were setting their employees up for disappointment. Employees don't want to be asked for feedback if they know nothing will be done with their opinions. A large number of employees will ignore the survey. The response rate typically runs about 30 percent in organizations where employees know that their feedback will not be acted on.

It would appear that a fundamental question needs to be asked by every organization: Why are we conducting an employee survey in the first place, and what are we going to do with the results?

From a strategic perspective, it seems reasonable to think that an organization would want, at the very least, to demonstrate that the survey is helping the organization to achieve their strategic goals. For example, the employee survey is a way of obtaining employee information that can be used to improve workplace practices in order to lift their employees' working experience. In turn, this will lift the customer experience and profits. However, if this or some other strategic purpose is not being fulfilled by the employee survey, then the value of conducting the survey is questionable.

Senior managers need to consider several points before embarking on an employee survey:

1. Are they committed to follow-up action?

2. Have they developed clear strategic objectives for the survey?

3. Have they developed a process for continuous improvement in which the survey plays a key data-gathering role and benchmarking role?

4. Will the survey questions provide the data needed to reach the survey objectives?

5. Will the survey analysis present the data in a way that will point clearly to follow-up action?

6. Has a pre-survey communication process been designed that will inform all the stakeholders (like unions) and employees of the objectives of the survey, address issues of privacy and confidentiality, senior management commitment to follow-up action, and post-survey communication?

7. Will management follow up with positive implementation?

8. When you compare your survey results to other companies, do you compare yourself to the best?

If this process is not followed, the organization can expect the following:

1. Employee participation rates in the survey will be low (30 percent or lower)

2. Employee cynicism with the organization will rise (why bother if the activity of completing an employee survey does not make a difference?)

3. Employees will become disengaged from the organization

4. The organization loses an opportunity to make significant strides in performance

Conclusion

The trap an organization falls into when they become focused on benchmarking themselves against others is that they lose sight of what is really important—what is it they are doing well, and where do they need to improve in order to create an even better organization than the one they already have? Focus on becoming better yourself and avoid the trap of comparing yourself to the industry average. If you must compare yourself to others, compare yourself only to the best and don't get sidetracked.

In Chapter 1, we separated out practices that take place at the team or department level, where local people managers have a reasonable amount of control over their work. This is distinct from the practices that have a broader corporate-wide reach that impacts most employees in the same way. These latter practices and policies come directly under the purview of senior management. The survey questions are grouped together along the lines of responsibility to show who in the organization has the authority to implement the follow-up. Our surveys are intentionally worded so that each question can lead directly to a follow-up action.

Over the years, we've read many employee surveys that were known by many different names: culture, climate, satisfaction, opinion, or engagement surveys.

The questions in all of these surveys were organized under various headings. The survey reports presented their statistics based on the headings, and no distinction was ever made in the survey report as to who had responsibility for follow-up action. Take communication, for example. The dynamics of communication at the departmental level are completely different than corporate communication. At the departmental level, local people managers have control over the nature of communication within their departments. Corporate communication is the responsibility of senior managers. Yet communication questions are typically lumped together and regarded in the same light. By not making this distinction, the survey results do not point to the group or individuals that have responsibility for follow-up action.

To confuse matters more, the surveys contain many vague questions that don't lead to follow-up implementation by specific individuals. They don't provide specific insights. The umbrella term "organization" is used frequently throughout these surveys. Some examples:

- Our organization motivates me to ...

- Our organization does an excellent job of ...

Many statements refer to the generic "managers," rather than the more specific "my manager." For example,

- The managers in our organization do a good job of ...

- Senior managers are effective leaders

All these examples beg the questions:

- Who, specifically, in the organization is motivating, or who is doing an excellent job?

- Which managers are doing a good job?

- What, specifically, are senior managers doing to make them effective leaders?

- Or what, specifically, are they not doing?

These surveys contain numerous questions that don't lead to a deeper understanding of the best practices that can be found in or are missing from the organization.

Respecting Individual Psychological Differences: The pitfall of time-based questions

Everyone is psychologically different. Therefore, the amount of positive feedback, recognition, etc., needed to boost motivation and self-esteem varies from one employee to another. Some employees are self-motivated and need very little feedback to work at peak performance. These employees will succeed no matter where they work. At the other end of the spectrum, there are employees who need much more attention. They constantly need feedback. Most surveys contain time-based questions that ignore this difference. For example:

How many times in the last three months:

- Did you receive positive recognition?
- Did your manager seek your opinion?
- Did you have contact with senior managers?

Suppose a person has not merited positive feedback during the past three months. Should they expect it? What is the relevance of having contact with a senior manager in the last three months? If my local people manager is doing an effective job of communicating, and if corporate communication is open and relevant, what useful purpose is being served by having contact with a senior manager every three months?

Clearly, there are exceptions. If a company is going through significant change brought on by a merger, a takeover, major financial problems, major restructuring, or significant growth, timely communication from a senior manager is essential. In these cases, a time frame of three months would not be adequate. Much more frequent communication is required. Time-based questions such as these provide little useful information and cannot be considered realistic measures of engagement. In fact, they can lead to misleading conclusions.

Points to Consider Before Conducting an In-house Employee Survey

For many years, I worked in the corporate environment in large organizations. These organizations had substantial human resource and information technology departments. When it came to running an employee survey, it was only natural that they should use these vast resources. Typically, response rates to employee surveys were less than 25 percent. This was considered the norm, and aside from a few comments lamenting the low response rate, no one paid much attention.

Today, I'm on the other side of the fence. I have worked with corporate clients for the past nine years, providing various kinds of employee surveys. Recently, I had the opportunity to speak with HR staff from several organizations. These organizations were scouring the Internet to collect questions so they could create their own employee survey and run the survey process in-house.

This might not sound like an unreasonable approach. However, I asked the HR staff to consider the following issues.

Privacy and Confidentiality

One of the most significant issues regarding employee surveys from an employee's perspective is privacy and confidentiality. It has been our experience that most employees have a low comfort level knowing that their responses to a survey are contained in their company's computers. Despite a company's best efforts to ensure that unauthorized access to the survey data is protected, the fact remains that it is company staff that are working with the data and conducting the analysis. Confidentiality has already been breached and there are many opportunities for abuses.

Several years ago, Entec was faced with a situation where the company president said he was prepared to move forward with an employee survey, but since he was paying for the survey, he wanted the database as well as the survey report we prepared. We had no choice but to walk away from the project, because we could not provide unequivocal assurances to the employees that their privacy and confidentiality would be secure.

These concerns can be significantly minimized when employees are advised that a third party will run the complete survey process. For example, in the pre-survey communications, Entec advises employees that there is a firewall between the organization and Entec Corporation. No employee can access our computers. No employee or company official will see or have access to our database. Any special requests by senior managers, or anyone else in the company, to look at the data is flatly rejected. (This has happened a couple of times over the past nine years.) Privacy and confidentiality are serious matters and they cannot be compromised. Employee surveys are a two-edged sword. On the one hand, employees welcome an opportunity to provide feedback. On the other hand, they will not participate or they will not provide honest answers if they feel there is the slightest chance their privacy can be compromised.

Response Rates

A high response rate generates a large database and will raise the statistical validity of the results. A large database can be used to prepare data cuts that drill deep into the organization, providing meaningful results. A small database can only be used to prepare a superficial analysis that will not be able to point directly to specific improvements that need to take place. Therefore, obtaining a high response rate is vitally important in any employee survey. The Entec survey process has resulted in response rates between 82 and 95 percent. This is well above average and it allows for detailed analysis. Data shows that company-run employee surveys typically garner a response rate of 30 percent or lower.

Survey Analysis

The nature of the survey analysis is just as important as the questions that are asked. The survey analysis involves more than providing percentages. The analysis must provide an interpretation of the statistics. For example, how do the answers from one question or set of questions relate to the answers to another question or set of questions? Some questions are much more important than others as they relate to employee motivation and performance.

In one company, the statement, "There is little to no office politics and gossip," statistically was linked with the following leadership statements: "Takes appropriate action with people who under-perform," "Resolves conflicts fairly and appropriately," and "Leads by example and action." This type of analysis identified other leadership behaviors that appeared over and over again as being behaviors that were important to the culture of the organization. The analysis led to identifying priority leadership behaviors that had the greatest impact on best practices. As a result, the HR department had a concise set of behaviors that needed to be coached on a priority basis. The survey report also provided an evaluation of how well all those in the company with supervisory responsibility rated against these behaviors.

These same statements don't necessarily link with the same behaviors in other organizations. They vary somewhat, depending on the company's culture. Considering the example of office politics, research has shown that a high level of office gossip is typically associated with a toxic workplace. This kind of analysis gives the company the knowledge and comfort that they are pursuing the right actions to minimize gossip and, therefore, improve performance.

There is also a particular ordering to the questions that drives the analysis, which in turn allows us to provide clear recommendations for follow-up implementation.

Reliability and Validity

Anyone can gather questions and create a survey, but it raises the question: How will they know that the questions are valid and reliable? If an organization creates its own survey, do they have the internal skills, and are they prepared to conduct the necessary reliability and validity testing to ensure the survey will produce meaningful results?

Nine years ago, Entec Corporation spent one whole year developing a series of surveys. The process involved a number of steps. The first step was assembling an eclectic group of professionals with expertise in strategic management, organizational development, leadership, psychiatry, and behavioral psychology. This group developed models and questions, based on these areas of expertise, that were tested with many focus groups in several business sectors, and then pilot tested. Reliability analyses and principle component analyses were conducted. The various surveys were pilot tested and analyzed again, and the surveys amended. This iterative process continued, and continues to this day, in order to ensure that clients receive surveys that will produce the best possible results.

Conclusions

If a person feels sick, is in pain, and is running a fever, they can do one of two things. They can either take their own temperature or they can go to the doctor. If they take their own temperature, their intervention options for regaining good health are severely limited because they don't have enough information. If they go to the doctor and undergo a battery of tests, they will receive valuable information and an intervention plan from their doctor.

Conducting an employee survey is the same. Organizations are complex human systems. Using an untested employee survey, untested survey process, and simple analysis will produce results that are similar to taking your own temperature. Taking your own temperature severely limits your ability to identify specific actions for performance improvement. This, in turn, acts as a de-motivator. Employee expectations are raised by employee surveys. When the post-survey process fails to show any meaningful movement, employee cynicism sets in and productivity drops.

Don't ask unless you mean it. Millions upon millions of dollars are wasted each year by companies that conduct employee surveys and then do nothing with the survey results afterwards. Questions are asked for which there are no useful answers and for which there can be no possible follow-up action. Employee response rates are low. Employee cynicism is high. But the greatest surprise comes from the fact that the surveys and the questions are approved by intelligent company presidents. This waste of money is being sanctioned by leaders who are otherwise closely watching their profit-and-loss statements. Employee cynicism and the resulting loss of productivity are being supported by senior managers who work hard to improve their competitiveness.

Employees as Customers: What HR needs to learn from Marketing

During the earlier stages of my career, I was fortunate to work for a large corporation that had a management development program that combined formal management courses with on-the-job training. The job training involved assignments to different divisions in the company. Two learning goals were mandated by these assignments:

1. Acquire knowledge in a new discipline

2. Learn about the different parts of the organization, experience their challenges, and understand how they contribute to the success of the whole

My formal education was in environmental studies with a specialty in ecology. One of the key principles in ecology is that ecosystems are made up of interdependent elements. A change in one part of an ecosystem will result in changes in other parts of the same system. Without knowing it at the time, my classmates and I became "systems thinkers." This ability to see systems has guided my decision making throughout my life, in business and in my private affairs. Naturally, I thrived as part of a management training program where I was able to experience different parts of the organization and see first-hand how each part related to the whole.

When I concluded the training program, I was appointed manager of marketing planning. My appointment coincided with a strategic decision made by the company to aggressively increase its share of the energy market. I had a staff of thirty-five and a budget of $3 million for market research. I was learning on the job. I learned from my staff, and I learned from the consultants we hired to conduct much of the market research. In addition, I was sent on a two-week intensive executive marketing program at the Graduate School of Business, Columbia University in New York, and a year later to the Wharton School of Business in Philadelphia. This was an amazing time of learning, personal growth, and achieving demanding goals.

After that, my career continued to flourish. I moved through the senior ranks of several companies until I reached president. Ten years ago, I established Entec Corporation, a company that specializes in measuring employee engagement. Although I loved marketing, I returned to my first passion—creating working environments where employees can thrive and be fully engaged.

Over the last ten years, I've worked with many organizations, and I've also been privy to the human resource practices of many others. To my surprise, I discovered that many HR departments lack research discipline when conducting employee surveys. In his article—"Is Job Happiness a Myth?"—Sudipta Dev of Aptech wrote about the importance of conducting an employee satisfaction survey as a way of gauging employee sentiment. He also mentioned how important it was to conduct focus groups afterwards to fully understand the survey results.

I've witnessed this process of conducting an employee survey, followed by focus groups, in several companies over the years, including a well-known company with thirty-five thousand employees. However, I thought these were just isolated cases. I was astonished to discover that this was quite common and considered a best practice. But it begs the question: Why spend money on an employee survey if it's going to be followed by focus groups? Isn't this doing research backwards?

Conducting an employee survey is conducting research. My marketing training and experience taught me that the survey is the last step, not the first step, in the research process. The purpose of the survey is to quantify and prioritize. Focus groups are used at the start of the research process to get an understanding of potential issues. In our marketing work, and now in Entec's HR work, we use the focus group information to develop a model first. This is followed by developing questions that fit within the parts of the model. Creating a model before developing the questions provides a framework for the questions. This framework provides a structure for the survey analysis, so that the results are organized and presented in a way that points clearly to follow-up action. When the survey and the analyses are completed, there is no question as to what the survey results mean. There is no question about priorities or about who is responsible for follow-up action.

ENERGIZING ORGANIZATIONS

Marketing and market research are sophisticated, disciplined processes that produce highly effective results. Automobile manufacturers use a variety of "focus group" techniques to clearly understand the reasons and motivators for a purchase decision: Is it external design, internal design, color, performance, quality, comfort, safety, size, fuel efficiency, financing, and so on? How will the different market segments prioritize these factors? The focus group information is used to develop the market research survey that will quantify the information. The research results are used to create the marketing programs for the various products and market sectors.

Employees are no less important than customers. Understanding the "root causes" of employee behavior and motivation is especially important in today's knowledge-based economy. We are in an economy where a company's success rests on the mental performance of its employees. It seems to me that in this environment, HR departments would bring greater value to their organizations if they adopted and applied marketing's sophistication and research discipline to understanding employee needs. A change in perception is required, where employees are viewed as customers in order to provide the information that unlocks their creative and innovative energy.

Let me share a personal story. Ten years ago, when Entec Corporation was founded, we spent the first year conducting research. The purpose of the research was to clearly understand the key factors that contributed to the employee experience in the workplace. We organized focus groups in several organizations from different business sectors. For example, the general manager of an electric utility consented to personally participate, along with half a dozen staff from different parts of his organization and different job levels. We facilitated many meetings over a three-month period to create an "employee experience model." The model depicted all the factors that contribute to the employee working experience. At the end of this period, the group formulated questions for an employee survey that was designed to measure the employee experience at work. The questions were clear and precise, and they led directly to follow-up action. This process was repeated at a healthcare facility and several other private-sector companies. The surveys were tested and validated.

When we used our employee survey, we noted that there was a direct link between the survey results and a company's financial performance. For example, we surveyed three electric utilities. Although the number of employees ranged from a hundred and fifty to four hundred, the customer profile for each utility was very similar. The revenue split between large industrial customers, commercial customers, and residential customers was about the same for each utility, so we were able to compare apples to apples. The utility with the highest employee survey scores was also the most profitable. The utility with the lowest employee survey scores was the least profitable.

Since that time, our employee models and surveys have evolved and become more sophisticated. Today we no longer talk about measuring the employee experience, but rather, we talk about employee engagement. When the employee surveys and analyses are completed, there is no question as to what they mean. There is no need for post-survey focus groups. There is a direct link between the survey results and the company's financial performance. The following note from a client summarizes this best:

> *Our company has partnered with Entec since 1999 to customize, implement, analyze and then action a compelling employee survey. I have reviewed and used many employee satisfaction instruments in the past, but none were as comprehensive, accurate or as linked to improving both business results and employee commitment as this one.*
>
> —Vice President Human Resources—International
> One of the largest retailers in the world

For companies and HR departments to view their employees as customers, they need to adopt the full spectrum of marketing concepts, processes, and tools to understand their employees and to meet their needs. This includes disciplined employee research, followed by appropriate communication, relationship building, and provision of products and services. Naturally, the products and services would depend on the survey results, but could include improved workplace practices, such as greater participation in decision making, infusing a high level of trust and fairness, choosing from a menu of benefits that best suit individual needs, consideration around work/life balance issues, zero tolerance policy on sexual harassment, verbal abuse and bullying, etc. Some companies are addressing many of these important issues, but frequently, the programs are developed in a piecemeal fashion. There is little knowledge about the value and contribution of each program to unlocking employee energy and to the bottom line.

A classic example of this is the company gym. I'm a great supporter of physical fitness. I exercise each morning. In the past, I worked for two companies that provided a physical fitness facility. I appreciated the convenience of these facilities; however, the fact that the facility was there did not change my behavior, and it did not seem to change the behavior of most other employees. Those who worked out did so whether there was a company gym or not. Those who did not exercise did not start exercising.

Typically, health departments measure the utilization rate of their gyms. But they don't measure relevant indices such as the "conversion rate" (i.e., the number of employees that did not exercise in the past but exercise now). They don't link the presence of a gym to the financial performance of the company. Is a gym the best way for a company to be spending its money? Should they be investing in strategically located meditation rooms, or a daycare center, or a full-time chaplain? Most companies cannot answer these questions because they don't have the information. They have not developed a framework to ask the right questions. They have not conducted disciplined employee market research.

Conclusions

In order to energize an organization to boost corporate performance, managers at all levels must create a working environment that will promote employee engagement and that will promote employee mental and physical health. At the same time, frontline employees and their immediate supervisors must take the initiative to improve those things over which they have control. They must lose the victim mentality and become full partners in improving their own working experience.

There are a number of valuable principles I hope the reader will take away from this book. An awareness of these principles is important to both managers and frontline workers. If managers take these principles to heart, it should equip them with the knowledge and tools to create working environments that will energize their employees and their organizations.

These principles are also important for frontline workers. All employees should be equally knowledgeable about the factors that affect their engagement and that affect their health. This understanding will help them to better understand why they may feel unmotivated, or frustrated, or tired and lethargic, rather than invigorated and energized by their work. Being able to identify and name a problem is the first step toward self-determination that leads to action.

The following is a brief summary of principles that impact employee engagement and health:

- Employee engagement is a partnership between the organization and the employees, where everyone works together to achieve the business objectives of the corporation and the personal aspirations of employees. Leaders at all levels of the organization have the responsibility to create the conditions for this to happen. Employees have the responsibility to willingly seek this partnership.

- Frontline supervisors and middle managers must be leaders. Leadership is not the preserve of senior management. Middle managers have a key responsibility to keep their employees fully informed about their organization's direction and performance. They have to do everything in their power to create the right working conditions that will allow their direct reports to be the very best that they can be. Local people leaders, in many respects, are the most important leaders in the organization. This is recognized by senior management in highly successful companies.

- In highly successful companies, senior managers assume full accountability for managing down their line. I have repeatedly witnessed directors and vice presidents completely preoccupied with politics and managing "up the line" in poor or average performing organizations, with little regard for actively managing "down their line."

- Department practices that have the greatest impact on employee engagement are those that unite employees around shared values: namely, fairness, justice, trust, and respect. In other words, all of the employees are working together in a cooperative community where they support each other, and from which they derive a strong sense of identity. Open communication and open sharing of information are key avenues through which the shared values are expressed and brought into being.

- Corporate practices that have the greatest impact on employee engagement are those that affect fairness of job promotions and opportunities for personal development. Related to that is an employee's desire to be recognized for a job well done. Salary is not a determinant of employee engagement, as long as the salary is considered to be reasonably normal within the context of industry standards. However, incentives and rewards for a job well done contribute significantly to employee engagement. Finally, open corporate communication about the direction and performance of the organization rank as the final significant contributor to employee engagement.

- A manager's personal characteristics of integrity and trust, and behaviors that show high standards of justice and fairness, play a greater role in employee engagement than do the behaviors that relate to job performance. In other words, employees will remain engaged if they work for a manager who exhibits high ethical standards, and who they trust, even though he or she may fall short on behaviors that are related to job performance, such as regular feedback and job appraisals.

- In the area of mission and values, the greatest impact on employee engagement is expressed by the focus on quality improvement, creativity, innovation, being clear and excited about the company mission in an atmosphere of kindness (no bullying or abuse), and where everyone behaves with integrity, honesty, and fairness. In many respects, the statements that have the greatest impact on employee engagement (in the section on mission and values) summarize the overarching elements of employee engagement: being excited about one's work, and being free to do a good job in a work environment that exhibits high standards of integrity, honesty, and fairness.

• The dynamics between employee engagement and employee health are complex. On the one hand, similar practices affect both employee engagement and employee health. On the other hand, different practices impact engagement and health. The values of quality improvement, creativity, innovation, being clear and excited about the company mission, in an atmosphere where everyone behaves with integrity, honesty, and fairness, had the greatest effect on employee mental and physical energy. These are the same as the organizational qualities that affect employee engagement. In addition, departmental practices dealing with open sharing of information, easy access to resources, and high levels of employee cooperation affected both employee engagement and employee mental and physical energy.

• Finally, corporate practices that affected employee engagement (incentives, fair promotions, opportunities for personal development, appropriate corporate communication, and appropriate health promotion programs) were the same as those that affected employee mental and physical energy. However, the degree to which an employee is able to contribute her abilities to her work each day, and the degree to which she is able to feel successful, affected her mental and physical energy more than her engagement. In addition, practices that are related to "time" (finding solutions in a time of crisis; completing work without the pressure of time; and being able to manage interruptions) link with mental and physical energy more strongly than with engagement.

Some Final Thoughts

1. Lack of leadership support, combined with poor work design, is a lethal combination that contributes to poor employee health and, in the extreme, can result in employee death.

2. An untrustworthy and unsupportive supervisor will demoralize otherwise motivated people.

3. In a toxic work environment, physical fitness and spiritual fitness cannot relieve stress over the long term. Eventually, the toxic work environment wins out.

4. As a middle manager in a large organization, you have the power to create a healthy and productive work environment if you have a vision, a plan, and the fortitude to see your plan through to a successful conclusion.

5. When creating change, identify two or three practices that you can get started on immediately that will yield the greatest results. All the other things that need changing will begin to fall into place.

6. Before starting a new job, ask your new boss about the worst possible situations that might be waiting for you. You will discover that they're all waiting for you, plus some you didn't expect.

7. As a frontline employee you have more power than you think to contribute to your working experience. Seize the opportunity.

8. Be a systems thinker. Study and follow the nature of connections and relationships between departments and managers. This will identify personal allegiances, partnerships, and biases. You will soon learn whose support you need or don't need to get your job done.

9. When it comes to values, the less you talk, the better. Let your actions speak for you.

10. Let your employees create your mission statement. You will be surprised; they know your business better than you do.

11. Keep your mission statement simple, so that your employees can carry it around in their hearts and minds.

12. Integrity rules. All business relationships—employees, suppliers, investors, community, and other stakeholders—must be based on the highest standards of integrity.

13. Recognize a bad working environment that cannot be changed, and get out sooner rather than later. You'll be glad you did.

Recommended Reading

Adrienne, Carol. *The Purpose of Your Life*. New York: William Morrow and Company, 1998.

Albrecht, Karl and Ron Zemke. *Service America*. New York: Dow Jones-Irwin, 1985.

_____. *Social Intelligence*. San Francisco: Jossey-Bass, 2006.

Barnett, Vicki. "Stress-related Disabilities Expected to Increase." *Calgary Herald*, May 8, 1999.

Bossidy, Larry and Ram Charam. *Execution*. New York: Crown Business, 2002.

Boothman, Nicholas. *How to Connect in Business in 90 Seconds or Less*. New York: Workman Publishing Company, Inc., 2002.

Buckingham, Marcus, and Curt Coffman. *First Break All the Rules*. New York: Simon & Schuster, 1999.

Burton, Terence T. and John W. Moran. *The Future Focused Organization*. New Jersey: Prentice Hall PTR, 1995.

Collins, Jim C. *Good to Great*. New York: HarperBusiness, 2001.

_____, and Jerry I. Porras. *Built to Last*. New York: HarperBusiness, 1994.

Cork, Tim. *Tapping the Iceberg*. Toronto: Bastian Books, 2007.

Covey, Stephen R. *The 7 Habits of Highly Effective People*. New York: Simon & Schuster, 1989.

_____. *First Things First*. New York: Fireside/Simon & Schuster, 1994.

DePaulo, J. Raymond and Leslie Alan Horvitz. *Understanding Depression*. New York: John Wiley & Sons, Inc., 2002.

DePree, Max. *Leadership Is an Art*. New York: Doubleday, 1989.

Foxman, Paul. *Dancing With Fear: Overcoming Anxiety in a World of Stress and Uncertainty*. Northvale, NJ: Jason Aronson, 1996.

Freudenberger, Herbert J. *Burnout—The High Cost of High Achievement*. Larden City: Anchor Press, 1980.

Giuliani, Rudolph W. *Leadership*. New York: Hyperion, 2002.

Goleman, Daniel. *Emotional Intelligence*. New York: Bantam, 1995.

Harvard Center for Population and Development Studies. *Global Burden of Disease and Injury Series*. (Harvard School of Public Health, 1993.)

Hemp, Paul. "Presenteeism: At Work—But Out of It." *Harvard Business Review* (October 2004): 49-58.

Jaworski, Joseph. *Synchronicity*. San Francisco: Berrett-Koehler Publishers, 1998.

Jones, Laurie Beth. *Jesus, CEO*. New York: Hyperion, 1995.

Kaplan, Robert S. and David P. Norton. *The Balanced Scorecard*. Boston, MA: Harvard Business School Press, 1996.

Kets de Vries, Manfred F. R. and Danny Miller. *The Neurotic Organization*. San Francisco: Jossey-Bass, 1984.

Kivimaki M, et al. "Organizational justice and health of employees: prospective cohort study." *Occupational and Environmental Medicine*, 2003; 60:27–34.

Kotter, John and James Heskett. *Corporate Culture and Performance*. Indianapolis, IN: Macmillan, 1992.

Likert, R. *The Human Organization: Its Management and Value*. New York: McGraw-Hill, 1967.

Maslach, Christina and Michael P. Leiter. *The Truth About Burnout: How Organizations Cause Personal Stress*. San Francisco: Jossey-Bass Publishers Inc., 1997.

McGregor, D. *The Human Side of Enterprise*. New York: McGraw-Hill, 1960.

_____, W. G. Bennis and C. McGregor (eds.) *Leadership and Motivation.* Cambridge. MA: MIT Press, 1966.

Murray, C. J. L., Lopez, A. D. (eds.) *The Global Burden of Disease: A Comprehensive Assessment of Mortality and Disability from Diseases, Injuries and Risk Factors in 1990 and Projected to 2020.* Cambridge Mass.: Harvard School of Public Health, 1996.

Ouchi, William. *Theory Z.* Reading: Addison Wesley, 1981.

Pascale, Richard Tanner and Anthony G. Athos. *The Art of Japanese Management.* New York: Warner Books, 1981.

Peck, Scott. *The Road Less Traveled.* New York: Simon & Schuster, 1978.

Pelletier, Kenneth. "A Review and Analysis of the Health and Cost-Effective Outcome Studies of Comprehensive Health Promotion and Disease Prevention Program at the Worksite: 1991-1993 Update." *American Journal of Health Promotion,* September/October 1993.

Pérez, Edgardo and Bill Wilkerson. *Mindsets.* Guelph: The Homewood Centre for Organizational Health at Riverslea, 1998.

Peters, Thomas J. and Robert H. Waterman, Jr. *In Search of Excellence.* New York: Harper & Row, 1982.

Pink, Daniel H. *A Whole New Mind.* New York: Penguin Group (USA) Inc., 2005.

Powers, W. T. *Making Sense of Behavior.* New Canaan, Conn.: Benchmark Publications, 1998.

Rutledge, Tim. *Getting Engaged.* Toronto: Mattanie Press, 2005.

Schaef, Anne Wilson and Diane Fassel. *The Addictive Organization.* New York: HarperCollins Publishers, 1988.

Schuster, John P., Jill Carpenter, and M. Patricia Kane. *The Power of Open Book Management.* New York: John Wiley & Sons, 1996.

Secretan, Lance H. K. *Reclaiming Higher Ground.* Toronto: Macmillan Canada, 1996.

Semler R. *Maverick.* New York: Tableturn Inc. Publishing, 1993.

Senge, Peter M. *The Fifth Discipline.* New York: Doubleday/Currency, 1990.

Siegrist, J. "Adverse health effects of high-effort/low-reward conditions." *Journal of Occupational Health Psychology,* 1996; 71:694–705

Smye, Mart and Anne McKague. *You Don't Change a Company by Memo.* Toronto: Key Porter Books Limited, 1994.

Stansfeld, S. A., et al. "Work characteristics predict psychiatric disorder: prospective results from the Whitehall II study." *Occupational and Environmental Medicine,* 1999; 56:302-7.

Tucker, Graham H. *The Faith-Work Connection.* Toronto: Anglican Book Centre, 1987.

Wang, J. "Perceived work stress and major depressive episodes in a population of employed Canadians over 18 years old." *Journal of Nervous and Mental Disease,* 2004; 192:160–3.

Wang, J. "Work stress as a risk factor for major depressive episode(s)." *Psychological Medicine,* 2005; 35:865–71.

Warren, Rick. *The Purpose Driven Life.* Grand Rapids: Zondervan, 2002.

Watson, Thomas J. *A Business and Its Beliefs.* New York: McGraw-Hill, 1963.

Welch, John F. and John A. Byrne. *Jack: Straight from the Gut.* New York: Warner Books Inc., 2001.

Wilkerson, Bill. "An Unheralded Crisis of Productivity." *Employee Health & Productivity,* October/November 1999.

"Worksite: 1991-1993 Update." *American Journal of Health Promotion,* September/October 1993, p51.

Ylipaavalniemi, J., M. Kivimaki, et al. "Psychosocial work characteristics and incidence of newly diagnosed depression: a prospective cohort study of three different models." *Soc Sci Med,* 2005 Jul; 61:111–22.

978-0-595-43185-4
0-595-43185-2

Printed in the United States
91360LV00006B/50/A